Geological Society of America
Memoir 187

Gondwana Master Basin of Peninsular India Between Tethys and the Interior of the Gondwanaland Province of Pangea

J. J. Veevers
Australian Plate Research Group
School of Earth Sciences
Macquarie University
North Ryde, N.S.W. 2109, Australia

and

R. C. Tewari
Department of Geology
D. S. College
Aligarh 202001
Uttar Pradesh, India

1995

Copyright © 1995, The Geological Society of America, Inc. (GSA). All rights reserved. GSA grants permission to individual scientists to make unlimited photocopies of one or more items from this volume for noncommercial purposes advancing science or education, including classroom use. Permission is granted to individuals to make photocopies of any item in this volume for other noncommercial, nonprofit purposes provided that the appropriate fee ($0.25 per page) is paid directly to the Copyright Clearance Center, 27 Congress Street, Salem, Massachusetts 01970, phone (508) 744-3350 (include title and ISBN when paying). Written permission is required from GSA for all other forms of capture or reproduction of any item in the volume including, but not limited to, all types of electronic or digital scanning or other digital or manual transformation of articles or any portion thereof, such as abstracts, into computer-readable and/or transmittable form for personal or corporate use, either noncommercial or commercial, for-profit or otherwise. Send permission requests to GSA Copyrights.

Copyright is not claimed on any material prepared wholly by government employees within the scope of their employment.

Published by The Geological Society of America, Inc.
3300 Penrose Place, P.O. Box 9140, Boulder, Colorado 80301

Printed in U.S.A.

GSA Books Science Editor Richard A. Hoppin

Library of Congress Cataloging-in-Publication Data
Veevers, J. J.
    Gondwana master basin of peninsular India between Tethys and the interior of the Gondwanaland province of Pangea / J.J. Veevers and R.C. Tewari.
       p. cm. -- (Memoir / Geological Society of America ; 187)
    Includes bibliographical references and index.
    ISBN 0-8137-1187-8
      1. Sedimentation and deposition--India.   2. Gondwana (Geology).
3. Basins (Geology).   4. Geology, Stratigraphic.   5. Geology--India.
I. Tewari, R. C.   II. Title.   III. Series: Memoir (Geological Society of America) ; 187.
QE571.G66   1995
551.7--dc20   95-32950
                                          CIP

10 9 8 7 6 5 4 3 2 1

Contents

Preface v

Abstract 1

Introduction 2

Pre-Gondwanan History 2

Succession of the Gondwana Master Basin 2
    Stages and Formations 3
        Talchir Formation 6
        Karharbari Formation 6
        Barakar Formation 6
        Barren Measures 7
        Raniganj Formation 7
        Panchet Formation 7
        Interval Between the Panchet and Supra-Panchet Formations 7
        Supra-Panchet Formation 7
        Triassic-Jurassic Formations in the Pranhita-Godavari and Son Areas 7
        Talchir-Panchet Column Beneath the Bengal Basin 7
    Structure 7
    Sedimentary-Tectonic Events 8

Paleogeographic Synthesis 8
    General 8
    Initial Relaxation of the Pangean Platform 9
    Palynologic Composition I or Talchir Stage 9
    Palynologic Compositions II and III or Karharbari and Barakar Stages 11
    Palynologic Composition IV or Barren Measures Stage 15
    Palynologic Composition V or Raniganj Stage 16
    Palynologic Composition VI or Panchet Stage 19
    Mid-Triassic (Anisian-Ladinian) Stage 19
    Carnian Stage 20
    Latest Triassic (Norian-Rhaetic) to Early Jurassic Stage 20

    Late Jurassic and Early Cretaceous Stage 21
    Late Cretaceous Stage 23
    K/T (Cretaceous/Tertiary) Stage 23

Connections with Tethys and Gondwanaland 32
    Global Scene at the 250-Ma Permian/Triassic Boundary 32
    Late Carboniferous (320 to 290 Ma) 33
    Carboniferous/Permian Boundary (290 to 275 Ma) 33
    Artinskian (265 Ma) 36
    Kungurian-Ufimian (260 Ma) 36
    Tatarian (253 Ma) 37
    Early Scythian (247 Ma) 41
    Late Triassic (230 to 208 Ma) 43
    Earliest Jurassic Deformation 44
    Early Cretaceous, Aptian (M0 = 118 Ma) 48

Connections Between India, Antarctica, Australia, and Africa 48
    Radial Drainage System about the Ancestral Gamburtsev Mountains 48
    Radial Drainage System Disrupted by Pangean Rifting 49
    Comparison of Damodar Basins with the Collie Basin 49

Pangean Tectonics and Stratigraphy 49
    (1) 320- to 290-Ma Lacuna and Glacials I 49
    (2) 290-Ma Pangean Extension I and Glacials II 52
    (3) Coal I 52
    (4) Coal Gap 52
    (5) Gondwanides II (~233 Ma) 52
    (6) 230-Ma (Carnian) Onset of Pangean Extension II and Coal II 54
    (7) Earliest Jurassic (208 Ma) Block-Faulting by Right-Lateral Transcurrence 54
    (8) Early Cretaceous Breakup of India from Antarctica and Australia 56

Acknowledgments 56

Appendix 1. Notes on Time-Correlation of the Gondwana Formations 57

References Cited 60

Index 69

Preface

The seed of this work was sown in 1985 by Tewari in a letter to P. J. Conaghan, Macquarie University, concerning the possibility of collaborative research and was developed in proposals from Veevers and Tewari to our respective national governments for a cooperative research project, "Comparative basin analysis of classical Gondwana (Indian) basins and equivalents in Australia." Veevers and Tewari joined forces first in April–June 1989 at Macquarie University, then in January and February 1990 at Aligarh Muslim University and in the field in the Damodar River area of Bihar and in the Umaria-Jabalpur (Son) and Pachmarhi (Satpura) areas of Madhya Pradesh, and finally in February 1994 at Macquarie University.

Between 1990 and 1994, Tewari filled a gap in information by measuring Triassic paleocurrents in the Gondwana basins. Other benefits of the long period of gestation came from the flow of information in the scientific papers of the Birbal Sahni Centenary National Symposium on Gondwana of India (Dutta and Sen, 1993) and in the Eighth (Hobart) and Ninth (Hyderabad) Gondwana symposia (Findlay et al., 1993; Mitra et al., 1994). Information about the Panthalassan margin of Gondwanaland came from "Permian-Triassic Pangean Basins and Foldbelts Along the Panthalassan Margin of Gondwanaland" (Veevers and Powell, 1994) and a Pangean perspective from Veevers (1988b, 1989, 1990a, 1994a). A final updating to mid-1994 of information outside the Gondwana area came from drilling in the Indian Ocean (e.g., Turner, 1991; von Rad et al., 1992b), Brookfield's (1993) summary of the Himalayan passive margin, Veevers and Tewari's (1995) account of Permian-Carboniferous and Permian-Triassic magmatism in the Indo-Australian rift zone, and Tewari and Veevers's (1993) and Veevers's (1994b) recognition of the ancestral Gamburtsev Subglacial Mountains of East Antarctica as the hub of the Permian-Triassic radial drainage in eastern Gondwanaland.

Any originality beyond Tewari's new fieldwork stems from the integration of Indian stratigraphic and tectonic events in a uniform timescale for Gondwanaland, as formulated in Veevers and Powell (1994) and in their setting between Tethys and the interior of Gondwanaland.

Gondwana Master Basin of Peninsular India Between Tethys and the Interior of the Gondwanaland Province of Pangea

ABSTRACT

Deposition in the Gondwana master basin of Peninsular India occurred during latest Carboniferous, Permian, Triassic, and Early Jurassic times on a basement of Archean and Proterozoic rocks between the Tethyan margin and interior of the Gondwanaland province of Pangea. Gondwana deposition ceased with the breakup of Greater India from the rest of Gondwanaland in the Late Jurassic and Early Cretaceous and was followed by a rift-drift succession along its margins.

Deposition in the Gondwana master basin started in the latest Carboniferous (Gzelian or 290 Ma) and continued through the Permian into the Early Triassic; after a Middle Triassic lacuna (except in the Godavari area), deposition resumed in the Late Triassic and terminated in the Early Jurassic. The master basin filled initially with lobes of glaciogenic sediment (Talchir Formation) in broad northwesterly valleys flexed between uplands generated by the first release of Pangea-induced heat. The end of the main glaciation in the Tastubian was accompanied by a marine transgression from the north, soon reversed by isostatic rebound. Coal measures were deposited in the coastal and fluvial valleys that subsided between growing normal faults—the Karharbari Formation by braided streams and the succeeding Barakar Formation by meandering streams. After a gap (Barren Measures), coal deposition resumed in the Late Permian Raniganj Formation from the meandering streams of valleys constrained by faulting in the Koel-Damodar area to flow to the west. Coal was lacking again in the Early and Middle Triassic succession of redbeds deposited by northwesterly flowing braided streams (Panchet Formation in the north, Kamthi to Bhimaram Formations in the south). Renewed faulting toward the end of the Middle Triassic was followed by Late Triassic deposition of the Supra-Panchet Formation in valleys renewed by Pangean rifting. Deposition ended during an Early Jurassic phase of intense transpression that dismembered the lobate master basin into individual structural basins in the Koel-Damodar area. The Late Jurassic and Early Cretaceous breakup of Greater India from the rest of Gondwanaland was accompanied by an onlapping rift-drift succession along the margins.

The Gondwana master basin lay 1,000 km inboard of the passive margin of Tethyan Gondwanaland in the distal part of a 10,000-km-wide radial drainage system that focused on a 2,000-km-distant upland in conjugate East Antarctica. In common with other distal parts of the drainage system in East Africa and Western Australia, the Gondwana master basin developed through the interplay of the Gondwanan climatic and biotal environment with the Pangean tectonics of Late Carboniferous initial subsidence and Late Triassic rifting of a Zambezian-type (anisotropic) basement followed by premonitory Early Jurassic internal dismemberment and definitive Late Jurassic and Early Cretaceous breakup.

Veevers, J. J., and Tewari, R. C., 1995, Gondwana Master Basin of Peninsular India Between Tethys and the Interior of the Gondwanaland Province of Pangea: Boulder, Colorado, Geological Society of America Memoir 187.

INTRODUCTION

The setting of Peninsular India within Gondwanaland is shown in Figure 1, with details in Table 1, and a map of the Gondwana structural basins is presented in Figure 2. "India" used alone is short for "Greater India" (Veevers et al., 1975) or the Indian subcontinent. The term "basin" denotes at least an area of sedimentary rock (as in "Satpura basin") and at most a body of sedimentary rock in a structural basin (as in "Jharia basin"). Each example is a remnant of the original depositional Gondwana master basin.

Elliot (1975), Mitra et al. (1979a), and Mitra (1993a) suggested that the Permian sediments of East Antarctica and Peninsular India were deposited in a master basin and that the Indian sediments were derived in part from Antarctica. We show that the Gondwana master basin was part of a 7,500-km-wide alluvial fan sourced from an upland in East Antarctica and that it was subsequently disrupted during stages of Pangean rifting and seafloor spreading.

PRE-GONDWANAN HISTORY

The structural grain of India was formed by Early and Middle Proterozoic mobile belts (EPMB and MPMB in Fig. 3) being wrapped around the Archean nuclei of Karnakata (KN), Jeypore-Bastar (JBN), and Singhbhum (SN) (Naqvi and Rogers, 1987). The resulting strongly anisotropic basement was prone to split along the grain in the same way as was the contiguous basement of the Zambezian-type (fault-affected) terrain of Africa (Rust, 1975). On this basement, the Late Proterozoic Vindhyan basin was deposited on a paleoslope (northerly arrows in Fig. 3) from the high ground southeast of the Narmada-Son lineament (N-S L), as much later, in the Permian and Triassic, the Gondwana master basin was deposited from the high ground to the southeast (Casshyap et al., 1993b). Following a later Proterozoic to pre–Late Carboniferous lacuna, the Pranhita-Godavari (PG) lobe of the late Paleozoic Gondwana master basin was deposited in a graben *parallel* to the trend of the mobile belt basement, whereas the Mahanadi (M), Son (S), and Rajmahal (R) basins and the Chintalapudi (C) and Krishna (KR) subbasins of the Pranhita-Godavari basin, southwest of the Mailaram Arch (M) (= Eastern Ghat-Sukinda-Singhbhum thrust), were deposited *across* the Proterozoic trend (Fig. 3). The Permian to Mesozoic boundary faults of the Koel (K)-Damodar (D) basins, some of which have a component of right-lateral displacement (Fig. 44 in a later section), trend at a low angle to the Proterozoic grain, consistent with initiation by incipient Early Permian right-lateral transtension.

SUCCESSION OF THE GONDWANA MASTER BASIN

The initial step in the analysis of the Gondwana master basin is to establish the biostratigraphic time-correlation of the constituent formations within and without the Gondwana region. We calibrate the biostratigraphic timescale in millions of years by the DNAG timescale (Palmer, 1983), modified such that the Permian-Triassic boundary is dated as 250 Ma, as in Veevers et al. (1994c). In aiming at a precise timetable of events, we had to rely on the stratigraphic distribution of the almost ubiquitous palynomorphs in the dominantly nonmarine Gondwana strata, supplemented by the rare Permian marine

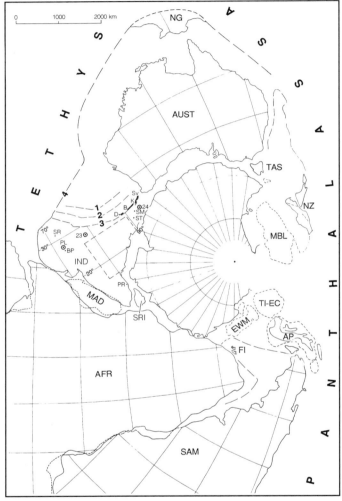

Figure 1. Setting of the Indian subcontinent within the Gondwanaland province of Pangea, between the Tethyan margin and the interior of Gondwanaland, bounded on the right by Panthalassa, on a base from Powell and Li (1994). The partial box shows the location of Figure 2. Details of localities in India are given in Table 1. The broken lines on the northeast side of India represent the Himalayan front (3), the Indus suture (2), the minimum probable northern extent of Greater India (1), and the maximum probable extent (4) (Johnson et al., 1980). The source of clasts in the Bap area (BP) of western India is the Proterozoic Malani rhyolite immediately to the south (Ranga-Rao et al., 1979). The encircled dot denotes information from drilling. AFR—Africa; AP—Antarctic Peninsula; AUST—Australia; EWM—Ellsworth–Whitmore Mountains; FI—Falkland Islands; IND—India; MAD—Madagascar; MBL—Marie Byrd Land; NG—New Guinea; NZ—New Zealand; SAM—South America; SRI—Sri Lanka; TAS—Tasmania; TI-EC—Thurston Island–Eights Coast; Sv—Siang Abor Volvanics.

TABLE 1. DATA IN FIGURE 1

Code	Locality	Formation	Reference
23	Bareilly	Talchir	Ranga-Rao et al., 1979, p. 485
24	Barpathar	Rangit	Srivastava et al., 1988, p. 329
B	Bhutan	Diuri Thungsing	Srivastava et al., 1988, p. 328
BP	Bap	Badhaura/Bap	Ranga-Rao et al., 1979, p. 485
D	Darjeeling	Upper Coal Measures Lower Coal Measures Rangit	Srivastava et al., 1988
K	Kameng	Bhareli Rangit	Srivastava et al., 1988, p. 328-329
PL	Pugal	Badhaura/Bap	Ranga-Rao et al., 1979, p. 485
PR	Palar	Talchir	Mitra et al., 1979, p. 36
S	Siang	Garu Rangit Abor Volcanics (v)	Srivastava et al., 1988, p. 329-331
SM	Singrimari	Talchir	Raja-Rao, 1981, plate 1
SR	Salt Range	Torba, etc.	Balme, 1970; Pakistan-Japanese Research Group, 1985
ST	Sylhet	Talchir	Srivastava et al., 1988, p. 329

invertebrates and Triassic land vertebrates. The correlation with the standard timescale and with the radiometric or numeric timescale given by Palmer (1983) allows us to cite ages in Ma, and the distinction between radiometric ages and biostratigraphic ages expressed in Ma is clear from the context. Comprehensive notes on the correlation of formations are given in Appendix 1.

We illustrate the succession of the Gondwana master basin with time-correlation diagrams (Figs. 4 to 6), columnar sections (Figs. 7, 9 to 14), facies diagrams (Fig. 8), cross sections (Fig. 15), a time-space fence diagram (Fig. 16), a thickness-variation diagram (Fig. 17), and a cumulative-subsidence diagram (Fig. 18). Supplementary information is given in Tables 1 to 4. The biostratigraphic correlation of the formations shown in Figures 4 to 6, 16, and 18 is documented in Appendix 1.

In brief, deposition in the Gondwana master basin started in the latest Carboniferous (Gzelian or 290 Ma) and continued through the Permian into the Early Triassic; after a Middle Triassic lacuna (except in the Godavari area), deposition resumed in the Late Triassic and terminated in the Early Jurassic. The breakup of India from the rest of Gondwanaland in the Late Jurassic and Early Cretaceous was reflected in a rift-drift succession that onlapped the margins.

Stages and formations

Pascoe (1968, p. 923–1016) suggested that the six main Permian to Early Triassic rock bodies or formations (Talchir Formation to Panchet Formation) were also stages (formations bounded by isochronous surfaces). Tiwari and Tripathi (1988) confirmed this suggestion by finding that each of the six formations contains a distinctive palynologic composition or assemblage zone (Table 2). Each composition is interpreted as reflecting a unique range of temperature and humidity so that over the small area of Peninsular India the change in composition is effectively isochronous. Palynologic composition I is coextensive with the Talchir Formation, which is therefore cognate with the Talchir Stage, as are the other succeeding compositions and formations to composition VI and the Panchet Formation/Stage (Figs. 4, 5). Stages for the rest of the Mesozoic are drawn from the general timescale (Fig. 6; Table 3).

The Talchir, Karharbari, and Barakar Formations/Stages extend into all the regions of the Gondwana basins, and their diagnostic features are given in Table 4. The overlying Barren Measures, Raniganj, and Panchet Formations/Stages are less extensive because of slight lateral facies variation—the main named variants are shown in Table 4. Of the three post–Early Triassic stages shown in Table 4, only the Supra-Panchet Formation has wide extent; most of the other formations are restricted to the Pranhita-Godavari region.

We interpret the uniform facies of the Talchir, Karharbari, and Barakar Formations/Stages as reflecting the continuity of each formation/stage within the Gondwana master basin and the increasing lateral facies variation that starts in the Barren Measures as reflecting increasing environmental differentiation within the master basin.

We summarize the facies and environments of the Gondwana succession from the work of Casshyap and Tewari (1984, 1988), augmented by other sources. The major formations in the Gondwana master basin are discussed in the ascending order of Talchir, Karharbari, Barakar, Barren Measures, Raniganj, Panchet, and Supra-Panchet (Mitra et al., 1979b; Mitra

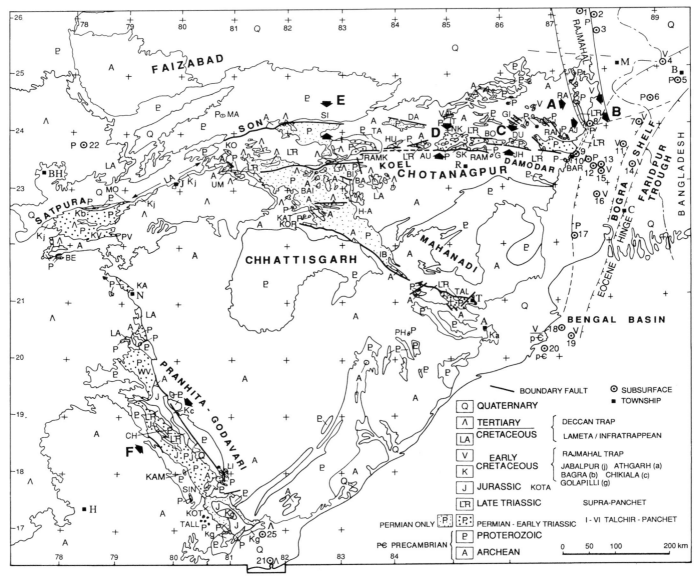

Figure 2. Gondwana basins of Peninsular India. Abbreviations explained in the key (facing). From Roy's (1962) Geological Map of India (scale 1:2,000,000), with these additions and changes: (1) Coalfields are from Fox (1934, pl. 14). (2) The marine or nonmarine Lameta Formation (LA) is mapped with the Infra-Trappean Beds (Robinson, 1967, p. 241). (3) Dikes west of Karanpura are not shown because Pascoe (1968, p. 1032) reported "Further to the westward no dykes are found in the coalfields until the area of the Deccan Trap is entered." (4) The boundary faults of the Koel-Damodar, Son, and Mahanadi basins are from Casshyap and Tewari (1984, fig. 1), Datta (1986b), and Singh (1988, fig. 6); faults in the Giridih area and to the east are from Casshyap and Tewari (1982, fig. 2) and Raja-Rao (1987, plate XVI); and faults at Itkhori are from Raja-Rao (1987, plate IX). (5) In the Barjora Coalfield, the Talchir Formation is unknown, and the 244-m-thick Barakar Formation, known mainly from drilling, rests on Archean basement (Narayana-Murthy and Hemmady, 1980). (6) The Cretaceous age of the Rajmahal and Bengal Traps is from Baksi et al. (1987, p. 134); (7) the map of the Jharia Coalfield is replaced by the map in Casshyap and Kumar (1987, fig. 1); (8) the North Karanpura basin is from Jowett (1929, plate 14); and (9) the Athgarh Formation, southeast of Talcher, is mapped from Kumar and Bhandari (1973, fig. 2). (10) In the Pranhita-Godavari basin, the faults are from Mishra et al. (1987), except the fault in the Chanda region that, following Pande (1988, p. 52), is removed; the Chintalapudi subbasin, south of 17.5°N, has been remapped by Lakshminarayana et al. (1991); southwest of the Chintalapudi subbasin, around Tallada, is a group of four outliers of Talchir Formation and to the southeast an outlier of Kamthi Formation (Raja-Rao, 1982, plate III; Mishra et al., 1987, fig. 1), which apparently rests direct on basement.

The arrow pairs A F indicate the location of the cross sections in Figure 15. The township northeast of Barjora (BAR) is Durgapur.

KEY TO FIGURE 2.

COALFIELDS

AJ	Ajai	HU	Hutar	KV	Kanhan Valley	RAN	Raniganj
AU	Auranga	IB	Ib River	LI	Lingala	SI	Singrauli
BA	Bansar	IT	Itkhori	MO	Mohpani	SIN	Singreni
BI	Bisrampur	JH	Jharia	NK	North Karanpura	SK	South Karanpura
BO	Bokaro	KA	Kamptee	PV	Pench Valley	TA	Tatapani
DA	Daltonganj	KO	Korar	RA	Rajmahal	TAL	Talcher
GI	Giridih	KOR	Korba	RAM	Ramgarh	UM	Umaria
H-A	Hasdo-Arand	KOT	Kothagudem	RAMK	Ramkola	WV	Wardha Valley

OUTLIERS OF PERMIAN-TRIASSIC ROCKS

NAME		LAT° N	LONG° E	REFERENCE
		DAMODAR BASIN		
G	Gola	23.55	85.77	Casshyap et al., 1993b; Mitra, 1993a
BAR	Barjora	23.40	87.30	Casshyap et al., 1993b; Mitra, 1993a
IT	Itkhori	24.15	85.00	Raja-Rao, 1987, pl IX
GI	Giridih	24.18	86.12	Raja-Rao, 1987, pl IX
AJ	Ajai	23.85	87.10	Fox, 1934
		SON BASIN		
BA	Bansar	23.05	83.40	Mitra, 1993a
MA	Maihar	24.30	80.75	Ranga-Rao et al., 1979, p. 485
		SATPURA BASIN		
22	Mirkheri	23.77	78.18	Singh, 1981
BE	Betul	21.80	77.80	Raja-Rao, 1983
		MAHANADI BASIN		
KOR	West of Korba	22.50	82.40	Raja-Rao, 1983
PH	Phulbani	20.40	84.30	Casshyap et al., 1993b; Mitra, 1993a
		PRANHITA-GODAVARI BASIN		
KAM	Kamaram	18.00	80.20	Casshyap et al., 1993b; Mitra, 1993a
TALL	Tallada	17.20	80.40	Casshyap et al., 1993b; Mitra, 1993a

INLIERS OF PRECAMBRIAN ROCKS

NAME		LAT° N	LONG° E	REFERENCE
		DAMODAR BASIN		
DU	Dumra	23.80	86.20	Casshyap, 1979a
		MAHANADI BASIN		
BAI	Baikunthapur group of 6	23.20	82.50	Raja-Rao, 1983
KAT	Katghora group of 4	22.70	82.50	Raja-Rao, 1983
		PRANHITA-GODAVARI BASIN		
CH	Chinnur	18.85	79.75	Kutty et al., 1988, p. 216

SUBSURFACE

NO.	NAME	OLDEST STRATA PENETRATED	REFERENCE
1	Purnea	Permian	Casshyap, 1979a, p. 537
2	A	Permian	Das et al., 1993
3	B	Permian	Das et al., 1993
4	Jaypur Hat	Early Cretaceous Vs	Alam et al., 1990
5	Kuchma	Permian	Alam et al., 1990
6	Singra-1	Permian	Neogi and Srivastava, 1987
7	Jalangi-1	Early Cretaceous Vs	Sengupta, 1966
8	Dewanganj	Permian	Saha et al., 1992
9	Kalipur	Early Triassic	Sengupta, 1966
10	Galsi-3	Permian	Sengupta, 1966
11	Debagram	Early Cretaceous Vs	Sengupta, 1966
12	C	Permian	Das et al., 1993
13	D	Permian	Das et al., 1993
14	West Ranaghat	Early Cretaceous Vs	Sengupta, 1966
15	Burdwan	Early Cretaceous Vs	Sengupta, 1966
16	Ghatal	Early Cretaceous Vs	Sengupta, 1966
17	E	Permian	Das et al., 1993
18	MND-1	Precambrian	Jaganathan et al., 1983
19	MND-2	Early Cretaceous Vs	Jaganathan et al., 1983
20	MND-3	Precambrian	Jaganathan et al., 1983
21	Narsapur	Deccan Trap	Jaeger et al., 1989
22	Mirkheri	Permian	Singh, 1981
23 (Fig. 1)	Bareilly	Precambrian	Ranga-Rao et al., 1979, p. 485
24 (Fig. 1)	Barpathar	Permian	Srivastava et al., 1988, p. 329
25	Kommugudem-A	Permian	Prasad and Jain, 1994, p. 240

SUBSURFACE COAL IN NORTHEAST

NO.	NAME	BASIN	THICKNESS (m) OF GONDWANA SEDIMENT	THICKNESS (m) OF COAL	% COAL
1	Purnea	Purnea sub-basin	1622	70	4.2
5	Kuchma	Bengal basin	498	110	22.0
6	Singra-1	Bengal basin	1200	200	16.6
10	Galsi-3	Bengal basin	834	112	13.6

TOWNS

A	Athgarh	C	Calcutta	M	Malda	R	Ranchi
B	Bogra	H	Hyderabad	N	Nagpur	T	Talcher
BH	Bhopal	J	Jabalpur				

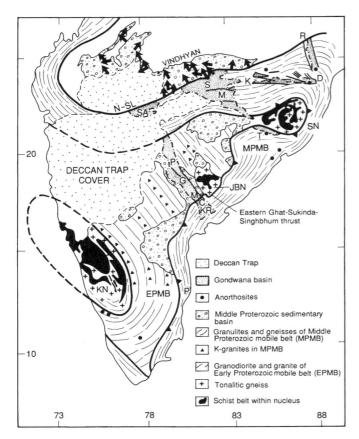

Figure 3. Geological map of Peninsular India, modified by permission from S. M. Naqvi and J. J. W. Rogers, Precambrian geology of India (Oxford: Clarendon Press, 1987) (fig. 1.4), showing component structural basins (stipple with dot-and-dashed outline) of the Gondwana master basin. This figure is the basis for Figures 20–30. Faults are denoted by the full heavy line. From north to south, the basins are Rajmahal (R), Koel (K), Damodar (D), Son (S), Mahanadi (M), Satpura (SA), Talcher (T), Pranhita (P), Godavari (G), Chintalapudi (C), and Krishna (KR). An outlier of the Talchir Formation is found at Palar (P) (latitude 13°N). The arrows indicate the paleoflow of the Middle-Late Proterozoic Vindhyan Supergroup (Casshyap et al., 1993b). Other abbreviations are: EPMB—Early Proterozoic Mobile Belt; JBN—Jeypore-Bastar nucleus; KN—Karnataka nucleus; M (south of JBN)—Mailaram Arch; MPMB—Middle Proterozoic Mobile Belt; N-SL—Narmada-Son lineament; SN—Singhbhum nucleus.

TABLE 2. PALYNOLOGIC COMPOSITIONS, FORMATIONS/STAGES, AND PALEOCLIMATE*

Composition	Formation/Stage	Temperature	Humidity
VI	Panchet	Cool-warm	Low to medium
V	Raniganj	Warm	Very high
IV	Kulti (Barren Measures)	Warm	Medium
III	Barakar	Cool-warm	High
II	Karharbari	Cold-very cold	Medium-high
I	Talchir	Extreme cold-cold	Low-medium

*Tiwari and Tripathi, 1988, figure 5.

and Raja-Rao, 1981). The following descriptions augment the diagnoses given in Table 4.

Talchir Formation. Unconformably or nonconformably overlying depressions in the Precambrian basement (Fig. 17 in a later section), the distinctly green Talchir Formation comprises locally massive but generally stratified tillite in thin beds associated with conglomerate, sandstone, interbedded fine sandstone-siltstone and shale (rhythmite), and shale (Fig. 7). Most clasts are derived locally from basement uplands, and some of those at the base are striated. Sandstone in the lower part is massive to thin bedded, and in the upper part it is cross bedded. In places, the rhythmite contains ripple- and flaser-bedding, in others dropstones; in others again, it contains turbidites deposited by underflow in lakes. Shale dominates downslope away from the basin margin and up-section above the basal tillite (Fig. 7). The marine section at Manendragarh (Fig. 7, VI B) lacks tillite and probably represents the youngest part of the formation.

Casshyap and Srivastava (1988) accounted for the variability of the Talchir Formation by suggesting two environments of deposition: glacial valleys in the southern uplands and a broad delta plain crossed by tidal channels in the low-lying northern Son-Mahanadi area (Fig. 8A).

Other sources of information include Bose et al. (1992), Sen and Banerji (1991), and Sen and Pradhan (1992).

Karharbari Formation. The Karharbari Formation gradationally overlies the glacial outwash of the Talchir Formation or overlaps it to rest on basement (Fig. 17). The Karharbari Formation consists of basal clast-supported conglomerate interpreted as braid bars of a low-sinuosity river, succeeded by multistory and multilateral coalescing channel bodies of pebbly coarse and medium sandstone, and at the top fining-upward cycles surmounted by coal (Figs. 8B, B′, 9).

Other sources of information include Niyogi (1961), Mitra et al. (1975), Tewari and Casshyap (1978, 1983), and Pandya (1990).

Barakar Formation. The Barakar Formation covers the previous formations to rest on basement (Fig. 17 in a later section). The Barakar Formation is a uniform set of fining-upward cycles of coarse to medium sandstone, interbedded fine sandstone or siltstone and carbonaceous shale, and coal (Figs. 8B, B″, 9). The sandstone is channel-shaped below and sheetlike above, with planar and trough cross-bedding, and is attributed to channel shifting and the lateral accretion of point bars. The interbedded sandstone and shale correspond to vertical accretion in levees and the coal to deposition in peat swamps in distal floodplains and lakes of meandering streams.

Other sources of information include Chowdhury (1990) on coccoliths, De (1990, 1993) on trace fossils, Aslam et al. (1991) on heavy minerals, Casshyap et al. (1988) on facies, Chakrabarti (1992) on coal thickness, Chakrvarti (1981) and

Ghosh and Mitra (1970a) on stratigraphy, Ghosh (1984) on sandstone fabric, Khan (1984) on grain-size analysis, and Mitra et al. (1975) on sandstone/shale ratios.

Barren Measures. The Barren Measures gradationally overlie the Barakar Formation and consist of repetitions of channel-like cross-bedded coarse to medium sandstone interbedded with siltstone and ironstone shale deposited in meandering streams (Fig. 8C and Fig. 10, column XXIII). Facies variants are the lower Pali Formation in the Umaria area and the Motur Formation in the Satpura and Kamptee areas (Table 4). The absence of coal and the local occurrence of phosphorite remain unexplained.

Raniganj Formation. The Raniganj Formation gradationally overlies the Barren Measures and is essentially a return to the style of deposition of the Barakar Formation (Figs. 8C, 10). Braided streams in the lower part give way to meandering streams, and in the Damodar area channel sinuosity increases downcurrent to the west (Fig. 8C). Phosphorite is found in the Raniganj area. Variants are the middle Pali Formation in the Umaria area, the lower Kamthi Formation in the northwest Pranhita-Godavari area, and the Bijori Formation in the Satpura area; an extreme variant, the Kamthi Formation in the Ib River area, is a set of barren redbeds. Other sources of information include Casshyap (1981a).

Panchet Formation. The Panchet Formation of the Koel-Damodar region conformably to disconformably overlies the Raniganj Formation or equivalents and consists of interbedded sandstone and shale (Figs. 8D, 11, 12, 13). It is distinguished from the underlying formations by its characteristic red and green shale, more micaceous and arkosic sandstone, and the absence of carbonaceous shale and coal (Sastry et al., 1977, p. 65). The Panchet Formation is interpreted as a bed-load deposit of braided streams (Fig. 8D). Variants (Table 4) are the upper Pali Formation in the Umaria area, the Pachmarhi Sandstone in the Satpura area (Fig. 11), and the middle and upper parts of the Kamthi Formation in the Kamptee and Pranhita-Godavari areas.

Other sources of information include Balasundarum et al. (1970) on redbeds.

Interval between the Panchet and Supra-Panchet Formations. The region that includes the Satpura, northwest Pranhita-Godavari, and Krishna-Godavari areas contains an interval of late Early and Middle Triassic formations that lie between the Panchet and Supra-Panchet Formations (Table 4). In the Satpura area, the Denwa Formation (Fig. 11) is a red clay with yellow sandstone that grades upward from the Pachmarhi Sandstone; in the northwest Pranhita-Godavari area (Fig. 14, column XXXVI), the Yerrapalli Formation and Bhimaram Sandstone are a succession of red clay and sandstone; and in Kommugudem-A well in the Krishna-Godavari area, unnamed sandstone and minor shale occupy this interval.

Supra-Panchet Formation. The Supra-Panchet Formation and equivalent Dubrajpur Formation unconformably overlie the older formations to rest on basement. They are distinguished from the underlying formations by containing pebbly sandstone (Figs. 12, 13) and are interpreted as a bed-load deposit of braided streams (Tewari, 1995). The equivalent Tiki Formation of the Umaria area and the Maleri Formation of the Pranhita-Godavari area consist of red and green mudstone and sandstone (Fig. 14), and the uppermost part of the unnamed formation in Kommugudem-A well in the Krishna-Godavari area consists of sandstone and minor shale.

Triassic-Jurassic formations in the Pranhita-Godavari and Son areas. In the northwest Pranhita-Godavari area, the Maleri Formation is disconformably overlain by the Norian Dharmaram Formation of pebbly sandstone and red mudstone, in turn overlain unconformably by the Early Jurassic Kota Formation, in turn overlain unconformably by the Cretaceous Chikiala Formation (Fig. 14). The Triassic and Jurassic formations are fluvial-lacustrine, and the Cretaceous formation fluvial. Further information is given by Bendapudi (1994) and Chandra (1991).

In the Son area, the Parsora Formation (Fig. 13), regarded here as Rhaetian but alternatively Middle Triassic (Fig. 5), consists of sandstone with mudstone mottled in violet and red.

Talchir-Panchet column beneath the Bengal basin. The 3-km-thick column (Fig. 14) northeast of the Damodar River and beneath the Cenozoic Bengal basin (#2 and #3 on Fig. 2) includes all the formations/stages from the Talchir through the Panchet.

Structure

Selected structures, shown in Figure 15, are (A) the successive westward overlap of the Barakar, Dubrajpur, and Rajmahal Formations over Archean basement (crosses) in the Rajmahal Hills; (B) the north-northwest– to south-southeast–trending boundary fault in the Dewanganj area that during event FA3 displaced the Talchir and Barakar Formations 700 m and is sealed by the Carnian Dubrajpur Formation; (C) the faulted Parbatpur dome in the Jharia basin; (D) the angular unconformity at the base of the Supra-Panchet Formation in the area of Rahum and Mandur in the North Karanpura basin, (E) the Singrauli coalfield of the Moher subbasin, with the 160-m-thick Jhingurdah Top Seam downfaulted against Proterozoic metamorphic rocks, (F) in the northwest Pranhita-Godavari basin, the boundary fault on the northeast, which started to grow (FA5) during deposition of the Early Jurassic Kota Formation and resumed movement (FA7) concomitantly with deposition of the Early Cretaceous Chikiala Formation (Fig. 6). The faults that expose the Proterozoic Sullavai Formation at Chinnur (between the 30- and 40-km marks) postdate the Carnian Maleri Formation and probably predate and accompany deposition of the Early Jurassic Kota Formation, which derived conglomerate from the uplift.

Structures elsewhere are shown in a later section (Fig. 44). Other sources of information include Ahmad (1966).

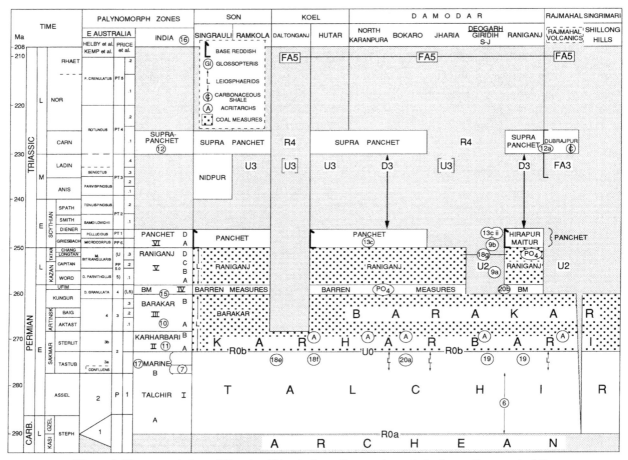

Figure 4. Time-correlation of Permian-Triassic formations and events in the northern part of the Gondwana master basin. Time scale from Palmer (1983); east Australian palynomorph zones from Helby et al. (1987), Kemp et al. (1977), and Price et al. (1985), as modified by Veevers et al. (1994e); and biostratigraphic correlation from Appendix 1. Tectonic and sedimentary events (e.g., R0a, FA5) are listed in Table 5. Other abbreviations and encircled numerals explained in text.

Sedimentary-tectonic events

The ages of structures are shown in Figures 4 to 6 and in Figures 16 to 18, which, together with the paleogeographic maps (Figs. 19 through 30), summarize the sedimentary-tectonic development of the Gondwana master basin.

The chief sedimentary-tectonic events (Table 5) include faulting, folding, relaxation by subsidence, rifting, coupled relaxation and uplift, uplift, and unspecified deformation. The distribution of these events through time and space is shown next on the paleogeographic maps.

PALEOGEOGRAPHIC SYNTHESIS

General

Time slices for the latest Carboniferous to earliest Triassic are those of the palynologic compositions or assemblage zones I to VI of Tiwari and Tripathi (1988), equivalent to the stages of Pascoe (1968, p. 923–1016) (Table 4). Stages for the rest of the Mesozoic are drawn from the general timescale.

According to Dutta and Suttner (1986) and Suttner and Dutta (1986, p. 331), "sediments [come] mainly from Precambrian granite and granite gneiss, subordinate amount of metasedimentary rocks." There was "little or no change in nature of parent rocks during the Gondwana sedimentation.... The maximum distance of transport by the streams supplying sediment to the basin was 100 km." Casshyap and Tewari (1984) draw a similar conclusion from the freshness of feldspar grains.

Suttner and Dutta (1986) distinguish *axial* meandering facies from *lateral* braided facies. This distinction is applicable to the Pranhita-Godavari and Mahanadi basins and to the Raniganj Formation in the Damodar basins but not to the Talchir and Karharbari-Barakar Formations in the Damodar basins, because the sediments crossed the east-west structural axis in a fan that radiated from an inferred source upland. The fan lay more than 1,000 km from the East Antarctic source, and the freshness of

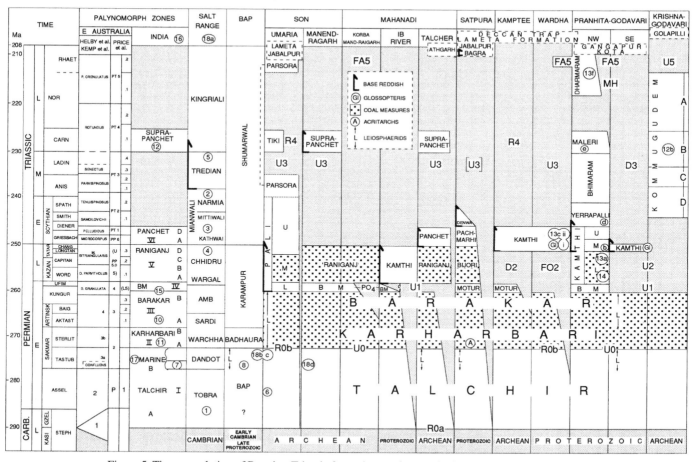

Figure 5. Time-correlation of Permian-Triassic formations and events in the central and southern parts of the Gondwana master basin. Sources as given for Figure 4. Biostratigraphic correlation from Appendix 1. Tectonic and sedimentary events (e.g., R0a, FA5) are listed in Table 5. Other abbreviations and encircled numerals explained in text.

the feldspar may owe more to the low temperature in the Permian than to proximity to the source.

In the following set of paleogeographic maps, we draw on the other synthetic figures—the time-correlation diagrams (Figs. 4 to 6), the fence diagrams (Figs. 16 and 17), and Table 5—to illustrate the geological development of the Gondwana basins of Peninsular India.

Initial relaxation of the Pangean platform

Initial relaxation of the Pangean platform near the end of the Carboniferous Period (290 Ma) provided broad depocenters for the accumulation of rock waste released from melting glaciers and ice sheets in glacially scoured depressions. By the Early Permian Tastubian time (or 275 Ma), deposits were widely distributed across Gondwanaland, including the Gondwana area inboard of what was to become the Triassic margin of Neotethys (Fig. 19, Table 6).

Palynologic composition I or Talchir Stage

Following initial relaxation (R0a in Table 5) of the platform of Peninsular India, the Talchir Formation, rock waste released from melting glaciers and ice sheets, accumulated in glacially scoured depressions (Table 7; Fig. 20). The eustatic rise in sea level during the Tastubian covered the northern part of Peninsular India until outpaced by postglacial isostatic rebound.

Studies of clasts in the Talchir Formation at Korba and till-fabric studies at Talcher indicate source uplands in the Proterozoic foldbelt basement of the Chhattisgarh upland. The upland extended northwest along the southwest flank of the Mahanadi-Son basin and west past Jabalpur to the east flank of the Satpura basin, as shown by the overlap of formations in the thickness fence diagram (Fig. 17). The northwest Pranhita-Godavari basin lay at the foot of an upland on the southwest. The Chotanagpur upland is indicated by glacial striations, crossdip azimuths, and till fabrics in the Koel-Damodar basins to the

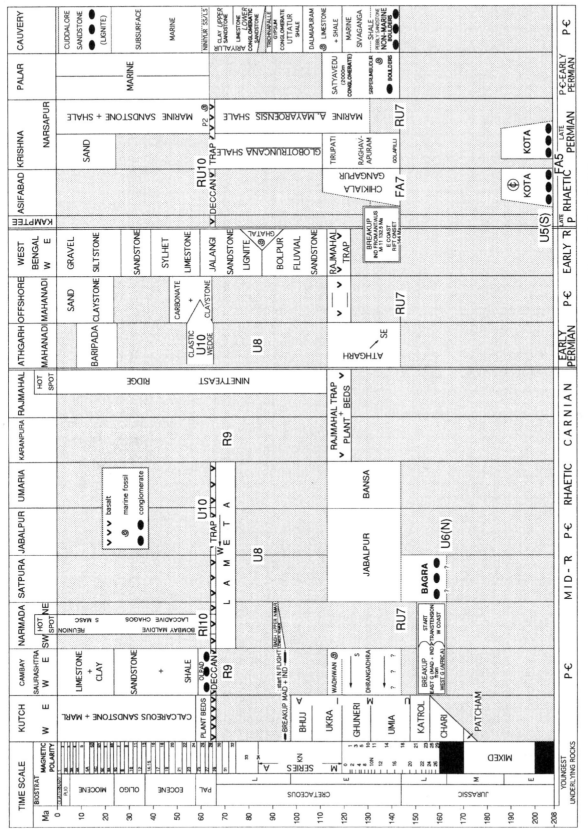

Figure 6. Time-correlation chart of Jurassic, Cretaceous, and Cenozoic formations of Peninsular India and adjoining areas, showing main tectonic and sedimentary events 5 to 10. Timescale from Palmer (1983), time-correlation from Bhalla (1983) except East Coast, from Sastri et al. (1973, 1974) and Zutsi and Prabhakar (1993). Source of details is given in Table 3, and tectonic and sedimentary events (e.g., U5, RU10) are listed in Table 5. From left to right are (1) areas in the north, from Kutch in the west to Rajmahal on the east; (2) the eastern part of the Mahanadi basin to West Bengal; and (3) the Pranhita-Godavari basin from Kamptee in the northwest to the Krishna-Narsapur area in the east, together with the other coastal basins at Palar and Cauvery in the south.

TABLE 3. SOURCE OF DATA IN FIGURE 6*

Column	Reference
Kutch	Biswas, 1971, 1981; Biswas and Deshpande, 1983; Krishna et al., 1983.
Cambay	Ramanathan, 1981; Raju and Srinivasan, 1983; Dhar and Singh, 1993.
Saurashtra	Biswas and Deshpande, 1983; Casshyap and Aslam, 1992.
Narmada	Jafar, 1982.
Réunion hot spot	Jaeger et al., 1989; Radhakrishna, 1989.
Satpura	Singh and Venkatachala, 1988; Baksi, 1994.
Jabalpur, Umaria	Brookfield and Sahni, 1987; Robinson, 1967; Singh, 1981.
Karanpura, Rajmahal, Ninetyeast Ridge-Amsterdam/St. Paul-Kerguelen hot spot	Baksi et al., 1987; Mahoney et al., 1983.
Athgarh/Mahanadi	Kumar and Bhandari, 1973.
Offshore Mahanadi	Jagannathan et al., 1983.
West Bengal	Roybarman, 1983; Neogi and Srivastava, 1987; Sen et al., 1987; Saha et al., 1992.
Kamptee, Asifabad	Rao and Yadagiri, 1981; Kutty et al., 1988.
Krishna-Narsapur	S. P. Kumar, 1983; Venkatachala and Sinha, 1986; Baksi et al., 1994.
Palar	Singh and Venkatachala, 1988.
Cauvery	Narayanan, 1977; Sastri et al., 1977; S. P. Kumar, 1983.

*Arranged in the order of the figure columns from left to right.

north and in the Son-Mahanadi basin to the west. The paleocurrents cross the Damodar basins; in the Son-Mahanadi basin, the paleocurrents are parallel to the trunk flow along the basin axis and normal to it by tributary flow. In the Pranhita-Godavari basin, the paleocurrents parallel the axis except in the Wardha region, where the ice-pavement striations trend east and northeast across the axis.

The incursion of the Tethyan sea from the northeast is indicated by the Tastubian marine invertebrate fauna (coils) at or near the top of the Talchir Formation at Umaria, Manendragarh, Daltonganj, Hutar, West Bokaro, and Ramgarh and additionally by leiosphaerids (L) in the Jharia and Rajmahal basins. Casshyap et al. (1993b) postulate a Faizabad ridge that divided the sea over the northern (Tethyan) margin into gulfs. Marking the edge of the Faizabad ridge are possible Talchir occurrences in the subsurface at Mirkheri (#22) and outcrops at Maihur (MA) and farther north at Bareilly in the Vindhyan Range (23 in Fig. 1). Past a presumed seaway between the Faizabad and Chhattisgarh uplands, leiosphaerids in the Satpura basin indicate the farthest southerly penetration of the Tethyan sea from the northwest, registered by marine invertebrates at Bap and the Salt Range (Fig. 1). This shoreline corresponds to the first-order Early Permian shoreline of Dickins and Shah (1981, p. 80). Leiosphaerids in the Palar basin near Mysore (Fig. 1) and in the southeast Pranhita-Godavari and Athgarh basins suggest the possibility of a marine incursion between India and Antarctica from the northern (Tethyan) margin through marine equivalents in the Perth and Carnarvon basins of Western Australia (Fig. 19).

The gross features of the Talchir paleogeography persisted into the postglacial Karharbari and Barakar Stages.

Palynologic compositions II and III or Karharbari and Barakar Stages

The Karharbari Formation consists of pebbly coarse sandstone, shale, and coal. The overlying Barakar Formation is finer and with only few exceptions lacks pebbles. It contains the chief productive coal measures and has been the principal subject of study by Gondwana geologists over the past 150 years.

Wide and fast subsidence (R0b, Table 5) led to the deposition of lobes of coal measures (Fig. 21) at the same time as the initial rifting in East Africa (Wopfner, 1993, fig. 2). Rejuvenated uplands (U0) were inherited from Talchir time, with the scarp of the Chotanagpur upland (Casshyap and Tewari, 1984, p. 138) indicated by the proximal facies of the Karharbari Formation (Mitra, 1987; Niyogi, 1987). The northern edge of the Chotanagpur upland was about 50 km from the Bokaro and Jharia coalfields. The Raniganj basin is the closest of the eastern Damodar basins to the postulated source region of the Chotanagpur upland, consistent with the occurrence at Maluncha Hill, in the southwest part of the Raniganj basin, of a conglomerate of subangular pebbles of gneiss and quartzite in the basal Barakar Formation that overlaps the Talchir Formation (Gee, 1932, p. 44).

The northern edge of the Chhattisgarh uplands was marked by a scarp in front of the proximal facies of the Korba coalfield. Pebbly sandstone in the Barakar Formation is notably thick in the Korba (100 m) and Talcher (60 m) coalfields (De, 1979). The Korba coalfield is close to the inferred constriction of the Mahanadi valley between the Chhattisgarh upland and the Chotanagpur upland; the basin widened to the southeast to accommodate the outlier at Phulbani (PH in Fig. 21).

The alluvial braidplain shed from the Chotanagpur upland radiated from northward in the Rajmahal area to northwestward in the Son area. Paleoflow in the Pranhita-Godavari-Satpura basin followed a simple axial pattern. The northerly paleocurrents at Umaria (Rishi, 1972) indicate a source in a northern extension of the Chhattisgarh upland now beneath the Deccan Trap.

TABLE 4. DIAGNOSTIC FEATURES OF GONDWANA STAGES/FORMATIONS*

Stage/Palynol. Composition	Formation	Region	Characteristics
———	Kota	Pranhita-Godavari	Top: limestone and calcareous shale; middle: red mudstone; base: sandstone with pebbles of banded chert.
	Dharmaram	NW Pranhita-Godavari	Alternating pebbly coarse crossbedded sandstone and red mudstone.
	Parsora	Umaria	Medium to coarse sandstone with micaceous mudstone mottled in violet and red; distinguished from Pali and Tiki Formations by presence of mottled mudstone and absence of feldspar.
Supra-Panchet	Supra-Panchet	All except Rajmahal, Umaria, Pranhita-Godavari, Krishna-Godavari	Pebbly to conglomeratic coarse sandstone with ferruginous siltstone and clay beds.
	Dubrajpur	Rajmahal	Same as Supra-Panchet Formation but with rare carbonaceous shales.
	Tiki	Umaria	Red mudstone and sandstone.
	Maleri	NW Pranhita-Godavari	Red and green mudstone with lenses of sandstone, in places distinctly white.
	Unnamed in Kommugudem-A	Krishna-Godavari	Sandstone, minor shale.
———	Unnamed in Kommugudem-A	Krishna-Godavari	Sandstone, minor shale
	Bhimaram Sandstone	NW Pranhita-Godavari	Coarse sandstone with red clay.
	Yerrapalli	NW Pranhita-Godavari	Red or green mudstone.
	Denwa	Satpura	Red clay and subordinate yellow sandstone.
Panchet / VI	Panchet	All except Umaria, Satpura, Kamptee, Wardha, Pranhita-Godavari	Interbedded sandstone and shale, distinguished from the underlying formations by its characteristic red and green shale, more micaceous and arkosic sandstone, and the absence of carbonaceous shale and coal.
	Upper Pali	Umaria	Interbedded chocolate to green shale and sandstone.
	Pachmarhi Sandstone	Satpura	Coarse white crossbedded sandstone with layers of pebbles.
	Upper Kamthi	Kamptee, Pranhita-Godavari	Coarse argillaceous sandstone with abundant quartz and quartzite pebbles in the upper part, and brick-red siltstone.
	Middle Kamthi	Kamptee, Pranhita-Godavari	Coarse argillaceous sandstone with clasts and lenses of purple siltstone.
Raniganj / V	Raniganj	All except Umaria, Ib River, Satpura, NW Pranhita-Godavari	A return to the style of deposition of the Barakar Formation: fining-upward cycles of coarse to medium sandstone, interbedded with fine sandstone or siltstone and carbonaceous shale, and coal.
	Middle Pali	Umaria	Green to reddish shale and white sandstone with interbedded coals.
	Kamthi	Ib River	Barren redbeds.
	Lower Kamthi	NW Pranhita-Godavari	Grayish-white calcareous sandstone and coal.
	Bijori	Satpura	Sandstone, carbonaceous shale, and coal.
Barren Measures / IV	Barren Measures	All except Umaria, Satpura, Kamptee	Repetitions of channel-shaped crossbedded coarse to medium sandstone interbedded with siltstone and ironstone shale; no coal.
	Lower Pali	Umaria	Red-brown clay and sandstone with carbonaceous shale.
	Motur	Satpura, Kamptee	Coarse sandstone with occasional clays and calcareous nodules.
Barakar / III	Barakar	All	Fining-upward cycles of coarse to medium sandstone interbedded with fine sandstone or siltstone and carbonaceous shale, and coal.
Karharbari / II	Karharbari	All	Top: fining-upward cycles surmounted by coal; middle: multistory and multilateral coalescing channel-shaped bodies of pebbly coarse and medium sandstone; base: clast-supported conglomerate.
Talchir / I	Talchir	All	Tillite associated with conglomerate and sandstone, interbedded with rhythmite (fine sandstone-siltstone and shale) and greenish shale. In places, the rhythmite contains ripple- and flaser-bedding, in others dropstones; in others again, it contains turbidites deposited by underflow in lakes.

*Tiwari and Tripathi, 1988, and Sastry et al., 1977, updated by references cited in the text.

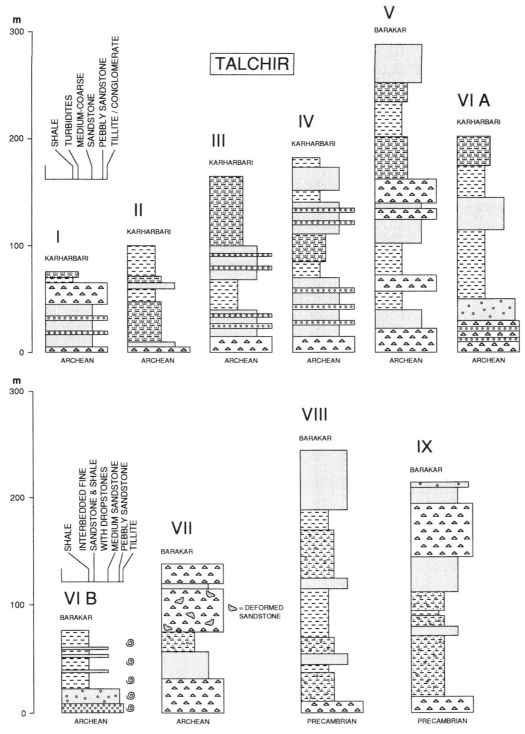

Figure 7. Representative stratigraphic columns of the Talchir Formation in the Koel-Damodar basin (I to VIA), Mahanadi basin (VIB), Pranhita-Godavari basin (VII, VIII), and Satpura basin (IX), showing lithology and grain size (top left). Thickness in meters. I (North Karanpura coalfield, located on Fig. 2), II (Ramgarh), III (East Bokaro), and IV (Jharia), all from Ghosh and Mitra (1975); V (Raniganj), from Banerjee (1966); VIA (Daltonganj), from Datta et al. (1979, p. 366); VIB (Manendragarh), in which coils indicate marine bivalves, from Casshyap and Srivastava (1988, fig. 13c); VII (central Pranhita-Godavari basin) and VIII (northwestern Pranhita-Godavari basin), both from Bose and Ramanamurthy (1979, p. 382); IX (Pench Valley coalfield), from Casshyap and Qidwai (1974).

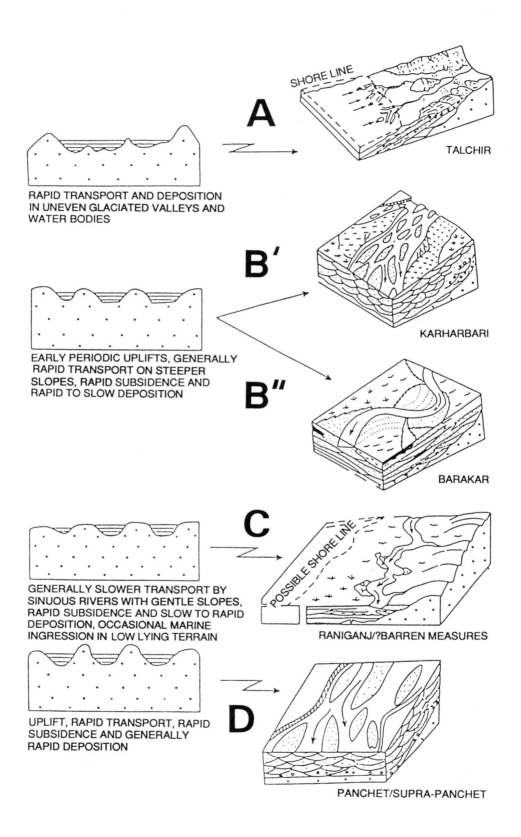

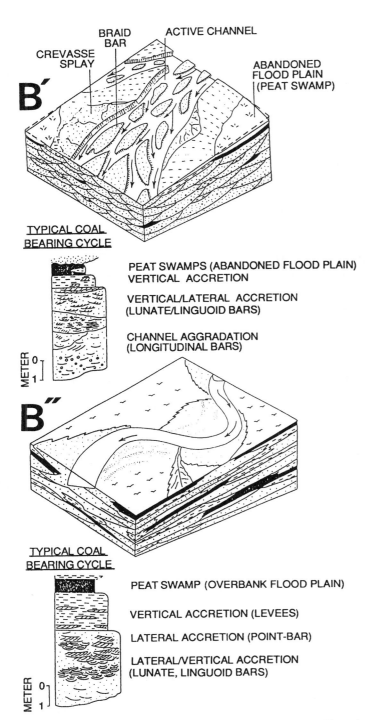

Figure 8 (on this and facing page). Block diagrams and profiles of depositional environments of (A) the Talchir Formation, (B) the Karharbari and Barakar Formations, (C) the Barren Measures and Raniganj Formation, and (D) the Panchet and Supra-Panchet ("Mahadeva") Formations, from Casshyap and Tewari (1988, fig. 2). Typical coal-bearing cycles and enlarged diagrams B′ of the Karharbari Formation and B″ of the Barakar Formation are from Casshyap and Tewari (1984, fig. 17).

A Monghyr-Saharsa ridge is postulated by Casshyap and Tewari (1984) and Casshyap et al. (1993b) to explain the divergent drainage on either side of a basement ridge in the western part of the Jharia coalfield. We suggest that the southeast cross-dip azimuth recorded by Khan and Casshyap (1979) (encircled 18 in Fig. 21) reflects this ridge.

A possible marine influence is indicated by leiosphaerids (L) and acritarchs (A). In an interpretation that regards these as definite indicators of a marine incursion over a low-lying coastal plain, we draw the maximum possible (and intermittent or ephemeral) penetrations of the shoreline from the Badhaura area of Rajasthan to the Satpura basin, thence northeast between the Faizabad and Chhattisgarh uplands to Umaria, into an indentation at Korba, then past Singrauli to Bokaro, and finally northeast across the Deogarh and Rajmahal basins to Singrimari and Subansiri. The total sulfur content in Barakar coal, highest at 1 to 2% in the Umaria coalfield (Tumuluri and Roychauduri, 1979, p. 237), is consistent with the postulated marine incursion in this area. High sulfur values in coal in the Warora coalfield (20.2°N, 79.0°E) are attributable to pyrite introduced from the Deccan Trap (H. K. Mishra, personal communication, 1990).

As shown by paleocurrent pattern, lithofacies, and paleohydrology (Casshyap and Tewari, 1984, 1988), the dominant depositional environment of the Karharbari and early Barakar Stages was braided rivers and of the late Barakar, meandering rivers. The trunk flow was axial in the Pranhita-Godavari basins. In the Koel-Damodar area, the flow was oblique to the present easterly trend of structure, but accumulation was greatest along the east-west axis, as shown by isopachs of the Barakar Formation and of individual coals and associated shales.

Palynologic composition IV or Barren Measures Stage

Facies range from nonmarine (lacustrine) and mixed marine-nonmarine (lagoonal or estuarine) mud ("Ironstone Shales") to nonmarine (meandering fluvial) sand and mud. The sandwiching of the Barren Measures between the Karharbari-Barakar and Raniganj coal measures reflects a pause in coal-forming conditions.

Again, in an interpretation that regards leiosphaerids and phosphorites as definite indicators of a marine incursion (Fig. 22), we draw the maximum possible (and intermittent or ephemeral) penetrations of the shoreline from the Salt Range to the Satpura basin, thence northeast between the Faizabad and Chhattisgarh uplands to Umaria, into an indentation at Ib River, then south of Singrauli to Bokaro, and finally northeast across the Rajmahal basin to Western Australia. The phosphoritic Barren Measures of the Karanpura and Bokaro basins lay along the lowlands of the northern coastal to shelf periphery of the fan in front of the Chotanagpur upland. Because the thickest Barren Measures in the Jharia and Raniganj basins lie on the present southern margin (Mitra, 1987, p. 33), the basin must have

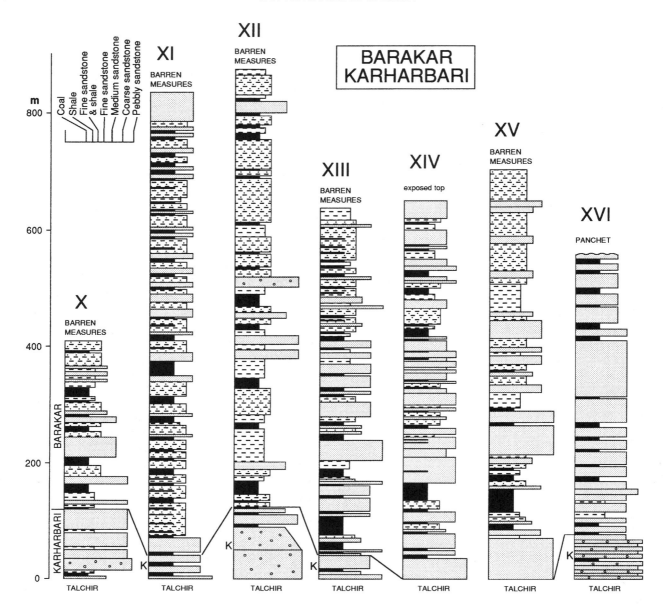

extended southward of the present southern boundary fault. The other phosphoritic Barren Measures, in the Korba and Ib River areas, together with associated leiosphaerids of the Ib River area, reflect a shoreline extending farther southeast from that of the Barakar Formation. Immediately southwest, the Talcher area was uplifted (U1) and eroded to become part of the Chotanagpur-Chhattisgarh upland.

In the exclusively nonmarine Pranhita-Godavari and Satpura basins, paleocurrents maintain a northwest axial flow downstream from the uplifted (U1) area in the southeast.

Scientific study of the Barren Measures has been neglected because they have been regarded as no more than a thick mineral band within the coal measures. It is hoped that the recent discovery of phosphorite (Datta, 1986a) will stimulate new work.

Palynologic composition V or Raniganj Stage

The Monghyr-Saharsa upland expanded westward and eastward across the northwest paleoslope of the Chotanagpur fan (Fig. 23) by uplift (U2) of the Rajmahal Hills and the region to the north and east, as indicated by the southwest paleocurrent into the Raniganj basin (encircled 25 on Fig. 23) (Casshyap and Tewari, 1984, p. 138, fig. 15), so that drainage was deflected westward to institute the Koel-Damodar valley (Casshyap and Kumar, 1987, p. 210, fig. 22). At the same time, the Chotanagpur upland expanded northward to the position of the present southern boundary fault of the Raniganj basin, indicated by debris-flow deposits at the base of the Raniganj Formation at Mehjia and Domra (solid triangle) (Mitra, 1987, p. 33). The

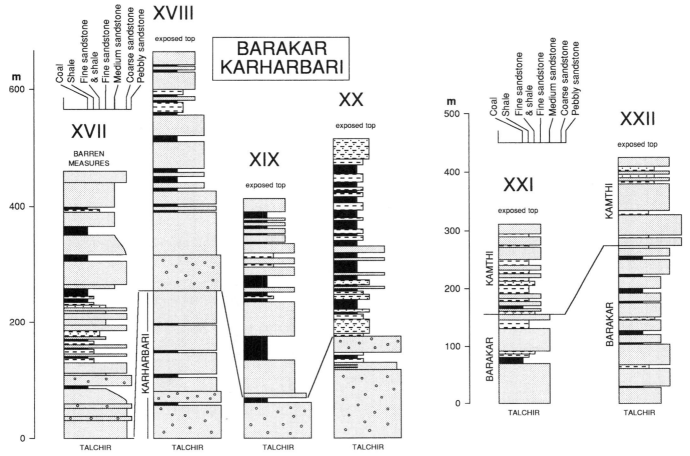

Figure 9 (on this and facing page). Representative stratigraphic columns of the Karharbari (K) and Barakar Formations in the Koel-Damodar basin (X–XIV, XVI), Rajmahal basin (XV), Son basin (XVII), and Mahanadi basin (XVIII–XX) and Barakar and Kamthi Formations in the Pranhita-Godavari basin (XXI, XXII), showing lithology and grain size. Thickness in meters. X (North Karanpura coalfield, located on Fig. 2), XI (South Karanpura), XII (East Bokaro), XIII (Jharia), XIV (Raniganj, Kulti area), XV (Chuperbhita), XVI (Raniganj, Shampur area), XVII (Singrauli), XVIII (Korba), XIX (Ib River), XX (Talcher), XXI (Wardha coalfield, located on Fig. 2), XXII (Ramagundam), all from Laskar (1979, p. 234).

Daltonganj–Deogarh–Rajmahal Hills terrain north of 24°N (north of the Damodar valley) was a nondepositional to erosional area from the start of Raniganj time. No Raniganj or later formations were deposited, and the previously deposited Talchir and Barakar Formations were eroded. In the Rajmahal Hills, sediment accumulated again in the Late Triassic (Dubrajpur Formation), and in the Early Cretaceous the whole terrain was covered by the Rajmahal Trap.

In the Wardha coalfield, an anticline grew during this stage (FO2, Fig. 16) until it was overlapped by the earliest Triassic Yerrapalli Formation. In the southeast Pranhita-Godavari area, the Kamthi Formation spread across the previous uplift (U1).

In the Talcher coalfield, the complete Raniganj Formation expanded across the previous uplift (U1) (Fig. 5).

Casshyap and Kumar (1987, p. 209) show that river sinuosity increases, sand beds get thinner, and finer grains increase down the axial paleoslope of the Damodar valley from the Raniganj basin westward. The marine outlet of the Damodar valley is possibly indicated by the leiosphaerid-bearing Raniganj Formation of the Singrauli Coalfield (encircled 24 in Fig. 23), which contains also the exceptionally thick, 134-m Jhingurdah Coal and dirt bands (Raja-Rao, 1983, p. 134) (Fig. 15, D). The inferred shoreline was the last sign of the Tethyan sea in Peninsular India. The only other known possible marine indicator is the phosphorite in the Raniganj area; in the absence of corroborative evidence, this occurrence of phosphorite in the highest known part of the valley is interpreted as lacustrine.

Another indication of a proximal source is provided by the few meters of conglomerate (with well-rounded clasts) in the North Karanpura Coalfield, supplied from a tributary stream with a west-northwest trend (Casshyap and Kumar, 1987, p. 185). The only direct evidence of the Monghyr-Saharsa upland is the southwest paleocurrent in the Raniganj basin, which we interpret as another tributary.

The Raniganj Stage contains coal of variable thickness:

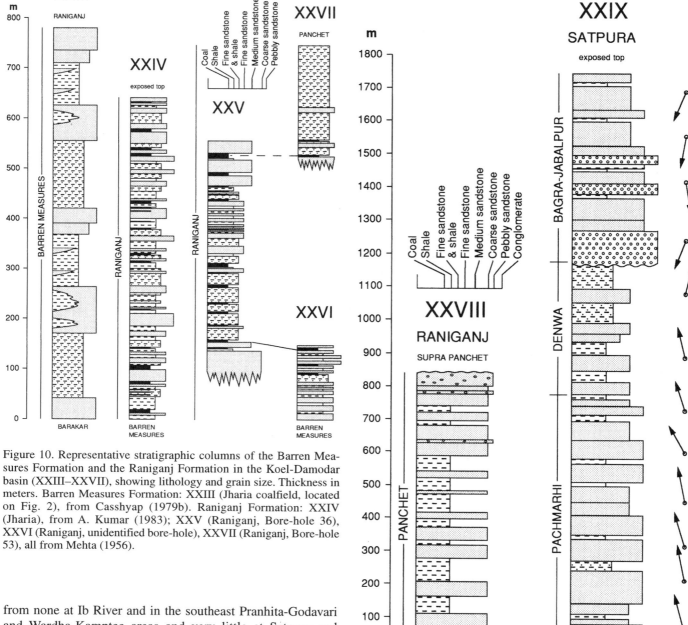

Figure 10. Representative stratigraphic columns of the Barren Measures Formation and the Raniganj Formation in the Koel-Damodar basin (XXIII–XXVII), showing lithology and grain size. Thickness in meters. Barren Measures Formation: XXIII (Jharia coalfield, located on Fig. 2), from Casshyap (1979b). Raniganj Formation: XXIV (Jharia), from A. Kumar (1983); XXV (Raniganj, Bore-hole 36), XXVI (Raniganj, unidentified bore-hole), XXVII (Raniganj, Bore-hole 53), all from Mehta (1956).

from none at Ib River and in the southeast Pranhita-Godavari and Wardha-Kamptee areas and very little at Satpura and Umaria to much elsewhere, with the extreme of the 134-m-thick dirty coal at Singrauli mentioned above. The bright red to purple pigmentation of the succeeding Panchet Stage is anticipated by isolated occurrences of redbeds in the lower Pali Formation of the Barren Measures Stage at Umaria, the Kamthi Formation of Raniganj Stage at Ib River, and the latest Permian middle part of the Kamthi Formation of the Kamptee and Pranhita-Godavari areas (Fig. 5). The low content of coal in the areas of pigmented Raniganj Stage suggests a drier environment than hitherto, in contrast to that of the Damodar-Singrauli valley, in which the favorable coal-forming conditions of the Karharbari-Barakar Stages returned.

Figure 11. Representative stratigraphic columns of the Panchet Formation (in the Koel-Damodar basin) and the Pachmarhi/Denwa/Bagra-Jabalpur Formations (Satpura basin), showing lithology, grain size, and paleoslope indicators. Thickness in meters. XXVIII (Raniganj coalfield, located on Fig. 2), from Casshyap (1979a). XXIX (Satpura basin), from Tewari and Casshyap (1994); the circle indicates the stratigraphic interval over which crossdip azimuths were measured and the arrow the vector mean. Note the 180° change in azimuth between the northerly crossdipping Denwa/Pachmarhi Formations and the southerly Jabalpur/Bagra Formations.

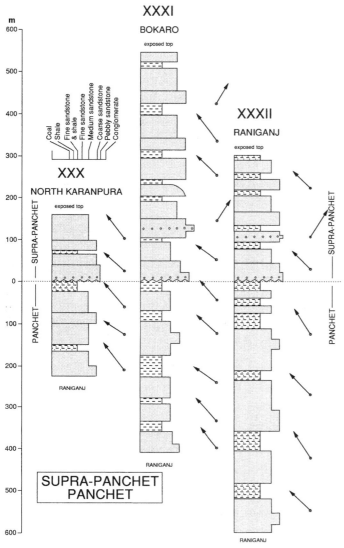

Figure 12. Representative stratigraphic columns of the Supra-Panchet/Panchet Formations in the Koel-Damodar basin, showing lithology, grain size, and paleoslope indicators. Thickness in meters. XXX (North Karanpura coalfield, located on Fig. 2), XXXI (Bokaro), XXXII (Raniganj), all from Tewari (1995). The circle indicates the stratigraphic interval over which crossdip azimuths were measured, and the arrows the vector means, which range from northwest to northeast.

Palynologic composition VI: Panchet Stage

During this stage, bright pigmentation and thin conglomerate layers appear everywhere, and coal is nowhere. This is the coal gap (Veevers et al., 1994b). Lacking coal and now commonly occupying higher scrub-covered terrain, the Panchet Stage did not attract comprehensive studies of paleocurrents until Tewari's recent work for this monograph. As shown on Figure 24, Tewari (1992) confirmed Dutta and Laha's (1979) undocumented northwest vectors in the Raniganj and Wardha areas and the north-northeast vector in the North Karanpura area and added new northwest vectors in the Damodar and Son areas; and Casshyap et al. (1993a) measured a northwest vector for the Pachmarhi Sandstone in the Satpura area, refining the north vector reported by Saxena (1963, p. 128). The Pachmarhi Sandstone is a braided stream deposit with pebbles of quartz and angular feldspar (Casshyap, 1979a, p. 531).

We interpret the paleogeography of the Panchet Stage as reverting to that before the Raniganj Stage, with paleodrainage to the northwest and north from the Chotanagpur upland—newly uplifted to provide a coarser grade of sediment—and confined between the Chotanagpur and Chhattisgarh uplands and within the Pranhita-Godavari valley. Following the relaxation of the Monghyr-Saharsa upland, the Chotanagpur fan spread to the north, and sediment (since lost) was probably deposited in the Deogarh-Rajmahal area.

Mid-Triassic (Anisian-Ladinian) Stage

Only four areas of deposition are preserved (Fig. 25). In the Satpura area (Raja-Rao, 1983, plate XV), the Denwa Formation is a meandering-stream deposit (Casshyap, 1979a, p. 531) of claystone with patches of conglomerate containing clasts of red jasper that are not found in the Pachmarhi Sandstone; vertebrates indicate an Anisian–early Ladinian age (240 to 233 Ma). The overlying Bagra Formation, formerly thought to be Triassic, is now dated as Jurassic, as detailed later. The second area of mid-Triassic rocks is the northwest Pranhita-Godavari basin (Raja-Rao, 1982, plate III; Kutty et al., 1988, map 1, p. 216). The Yerrapalli Formation is a meandering stream deposit of red and purple clay with calcareous sandstone at the top (Dasgupta, 1993); paleocurrents are to the north. The Bhimaram Sandstone is a braided river deposit of coarse to pebbly sandstone with red clay intercalations; paleocurrents are to the northwest. We interpret the upward coarsening from the Yerrapalli to the Bhimaram as reflecting a rejuvenation by uplift of the source areas. The third area, to the southeast, is the Krishna-Godavari area, with the shale and sandstone penetrated in Oil and Natural Gas Commission (ONGC) Kommagudem-A well (Prasad and Jain, 1994). The fourth occurrence, in the Son area, is an isolated outcrop of shale and sandstone, the Nidpur Beds, that contains Anisian-Ladinian plants.

Deformation (D3) and uplift (U3) toward the end of the mid-Triassic (233 to 230 Ma) are recorded from all areas that contain Late Triassic rocks, except the northwest Pranhita-Godavari area, which contains a continuous record of deposition over the interval. The style of deformation in the Rajmahal area is indicated by a north-northwest–trending normal (extensional) fault (FA3) sealed by the Late Triassic Dubrajpur Formation (Fig. 15, A). Other faults are sealed by the Supra-Panchet Formation in the North Karanpura area (Fig. 15, D), and most of the deformation was probably effected by normal, extensional faulting. We interpret the main northerly downthrow on the line of southern boundary faults of the Koel-Damodar area and its western continuation across the Son basin as happening at this

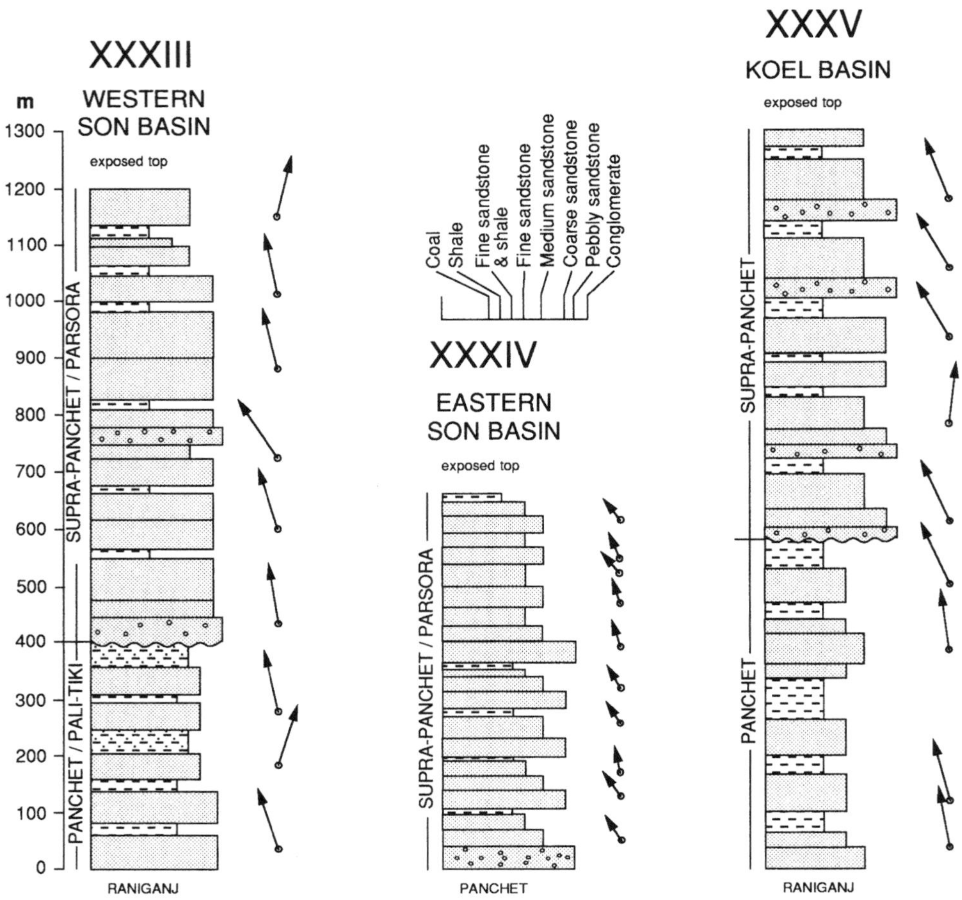

Figure 13. Representative stratigraphic columns of the Panchet/Pali-Tiki, Supra-Panchet/Parsora, and Panchet/Supra-Panchet Formations, showing lithology, grain size, and paleoslope indicators. Thickness in meters. XXXIII (western Son basin, located on Fig. 2), from Tewari and Casshyap (1994); XXXIV (eastern Son basin), from Tewari (1995); XXXV (Koel basin), from Tewari and Casshyap (1994). The circle indicates the stratigraphic interval over which crossdip azimuths were measured and the arrow the vector mean, which ranges narrowly about the north.

time because the Carnian Supra-Panchet Formation lies wholly north of this line.

Deformation 3 (D3) coincides with the Gondwanides II deformation along the Panthalassan margin of Gondwanaland and the Indo-Sinian orogeny (Veevers, 1989; Veevers et al., 1994c).

Carnian Stage

Deposition continued in the northwest Pranhita-Godavari area and the Krishna-Godavari area and resumed in the Talcher area and, during relaxation (R4) after the mid-Triassic deformation and downthrowing, in the northern Son area and Koel-Damodar area (Fig. 26). In the Koel-Damodar area, the Supra-Panchet Formation unconformably overlies the Panchet Formation, and in the Rajmahal area, the Dubrajpur Formation unconformably overlaps the faulted Talchir and Barakar Formations to rest on Precambrian rocks (Fig. 15, A). The Supra-Panchet comprises reddish coarse sandstone and thin shale interbeds, and the slightly coarser Dubrajpur Formation contains conglomerate, sandstone, shale, and thin coal bands (¢), the latter unique in the Indian Triassic. Both formations are braided river deposits. The paleocurrent vectors indicate a northerly paleoslope that repeats the paleogeography of the Panchet and Barakar Stages. In the Son area, the Supra-Panchet contains northwest paleocurrents that indicate a paleoslope along the axis of the Mahanadi basin. In the northwest Pranhita-Godavari basin, the Maleri Formation of red-green clay with lenses of sandstone and lime-pellet rocks is a meandering-river deposit with a northerly paleocurrent. The rocks in the Talcher area lack fossils but are regarded as equivalent from their similarity with the Supra-Panchet Formation.

Latest Triassic (Norian-Rhaetic) to Early Jurassic Stage

Outcrops are restricted to the Pranhita-Godavari basin (Dharmaram and Kota Formations) and possibly the Son-Umaria area (the poorly dated Parsora Formation) (Fig. 27). The Dharmaram Formation is a fluvial deposit in which coarse to pebbly sandstone alternates with clay bands. The Kota Formation is another fluvial deposit of pebbly sandstone with clay bands and thin coal seams and at the top limestone that contains a fish fauna (Bhattacharya, 1981). Marly beds underlying the limestone contain vertebrates (Dutta and Yadagiri, 1994). The coarse grade of the sandstone is interpreted as indicating intermittent uplift of the source area (Rudra, 1982, p. 77). Mitra

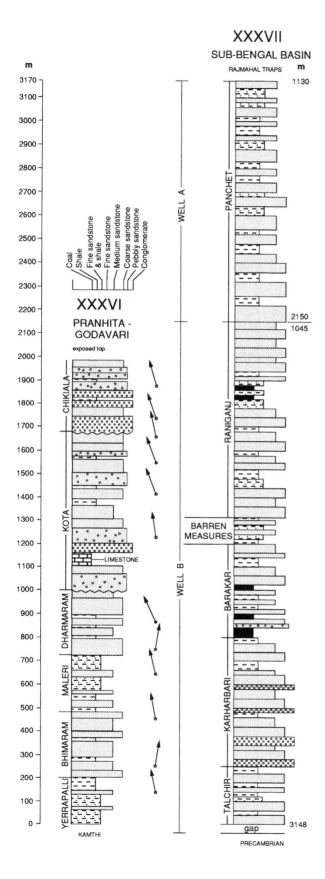

(1987, p. 36) interprets the basal pebbly sandstone as a debris-flow deposit that indicates activation of the northeast border fault in the Early Jurassic. We interpret further that the faults that expose the Chinnur inlier became active at this time. Paleocurrents indicate a continuation of the northwest axial slope away from the newly formed horst of the Mailaram High. The Parsora Formation of the Son-Umaria area is medium to coarse to pebbly sandstone interspersed with mudstone. As discussed earlier (Fig. 5), it is dated as mid-Triassic or, as we show here, Rhaetian. Its northwest paleocurrent may reflect the transpressional uplift to the east.

At some point within the interval between the Carnian Supra-Panchet/Dubrajpur Formations and the Early Cretaceous Rajmahal Trap/lamprophyre, the Damodar basins were deformed and uplifted by transpressional motion on the boundary faults and northwest-trending cross faults. We regard this event as the same age (FA5) as the faulting in the northeast Pranhita-Godavari area, dated as immediately before and during deposition of the Early Jurassic Kota Formation. Mitra (1987, fig. 2) shows right-lateral movement along the boundary faults of the Damodar area, and Veevers et al. (1994a, fig. 3) illustrated the 4.4 km of right-lateral offset, as shown in Figure 44 in a later section. They also analyzed the structure of the Jharia basin as showing right-lateral movement. Mitra (1987, fig. 3) illustrated the right-lateral transpressional regime superimposed on the earlier extensional structures to generate uplift in the Son Valley area, and this is shown on Figure 27 by the parallelogram labeled FA5.

Late Jurassic and Early Cretaceous Stage

During the Late Jurassic and Early Cretaceous, India broke off Africa on the southwest and Antarctica on the southeast. Deposits include the (?) Late Jurassic Bagra Formation in the Satpura area, the Early Cretaceous Jabalpur Formation of the Satpura-Jabalpur region, the Athgarh and Golapilli Sandstones of the east coast (Rao et al., 1993), the Chikiala- Gangapur Formations of the northwest Pranhita-Godavari basin, and the

Figure 14. Representative stratigraphic columns of Mesozoic formations in the Pranhita-Godavari basin and of Permian and Triassic formations in the eastern subsurface Koel-Damodar basin beneath the Cenozoic Bengal basin, showing lithology, grain size, and paleoslope indicators. Thickness in meters. XXXVI, Triassic Maleri Formation, Early Jurassic Kota Formation, and Early Cretaceous Chikiala Formation (central Pranhita-Godavari basin, located on Fig. 2), from Tewari and Casshyap (1994); the circle indicates the stratigraphic interval over which crossdip azimuths were measured and the arrows the northerly vector mean. XXXVII, Talchir through Panchet Formations in the eastern subsurface Koel-Damodar basin, beneath the Cenozoic Bengal basin, from Das et al. (1993, fig. 3); the column was constructed from Well A (interval 1,130 to 2,150 m) and Well B (1,045 to 3,148 m), which terminated in the Talchir Formation, a short interval above the Precambrian basement. The wells are located at #2 and 3 on Figure 2.

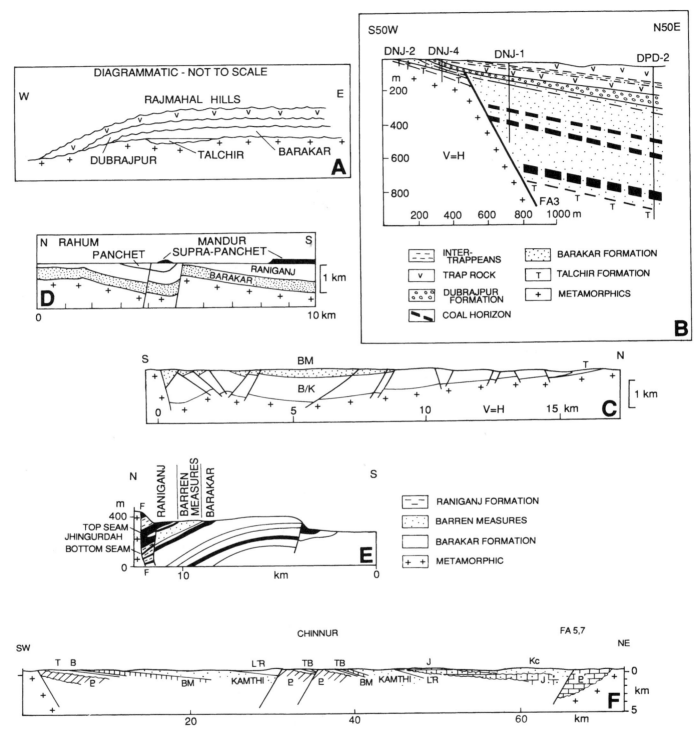

Figure 15. Cross sections selected to show principal structure, located on Figure 2. A, Diagrammatic section of the Rajmahal Hills. Drawn from information in Raja-Rao (1987). B, Cross section of the Dewanganj area (#8 on Fig. 2), from Sen et al. (1987, fig. 2). DNJ-1, etc., are drill-holes. C, Jharia basin, from Ghosh and Mukhopadhyay (1985, fig. 4B). BM—Barren Measures (stipple); B/K—Barakar Formation over Karharbari Formation; T—Talchir Formation. D, North Karanpura basin in the area of Rahum and Mandur. From Jowett (1929, plate 13, section A). E, Singrauli coalfield. From Dutta and Mukherjee (1979, plate 121). F, Northwest Pranhita-Godavari basin, Chinnur Inlier. From Raja-Rao (1982, plate III). Abbreviations, in stratigraphic order: crosses—Archean rocks; P—Proterozoic rocks; T—Talchir Formation; B—Barakar Formation; BM—Barren Measures Formation; LR—Late Triassic Maleri Formation; J—Jurassic Kota Formation; Kc—Cretaceous Chikiala Formation.

Rajmahal Trap and lamprophyre and dolerite dykes and sills in the Damodar, Deogarh, and subsurface Bengal Basins and the offshore Mahanadi basin (Fig. 28), all dated as 120 to 110 Ma, within the Early Cretaceous. The sedimentary formations are piedmont fluvial deposits.

The Bagra Formation is an alluvial fan deposit (Casshyap, 1979a, p. 531; Casshyap et al., 1993a) that contains pebbles to head-sized boulders of quartzite, matched with the Proterozoic Vindhyans from Hoshangapad immediately to the north and northeast, and banded jasper and jasperoid conglomerate, matched with the Proterozoic Bijawar and Kaimur Formations, again north and northeast of the Narmada valley (Pascoe, 1968, p. 970). The Bagra Formation overlaps the exposed Precambrian rocks on the northern edge of the Satpura basin and then progrades southward over the Denwa Formation down a paleoslope reversed from that of the Denwa Formation and Pachmarhi Sandstone. This reversal of paleoslope signifies a major uplift in the immediate north, the Hoshangabad upland (U6) with its eastern extension, the Vindhyan upland (U7), across the earlier northward drainage.

Paleocurrents in the Early Cretaceous Jabalpur Formation indicate drainage to the south-southwest; the Jabalpur Formation overlaps the Precambrian basement to indicate the demise of the northwest extension of the Chhattisgarh upland.

The Chikiala Formation contains a piedmont breccia-conglomerate deposited at the foot of the boundary fault along the northeast margin of the northwest Pranhita-Godavari basin (Fig. 15, F). The adjacent Gangapur Formation overlies the Kota and Maleri Formations "with a clear angular unconformity" (Sastry et al., 1977, p. 37). Both formations indicate movement (FA7) on the boundary fault during the Early Cretaceous.

The southeast paleocurrents in the coastal Gondwanas mark a reversal of paleoslope from the northwest paleoslope of the Permian to Early Jurassic Pranhita-Godavari basin (Golapilli Formation) and the Early Permian Mahanadi basin (Athgarh Formation). The reversal is attributable to the thermal subsidence of the eastern margin (Casshyap and Tewari, 1988, p. 65) after its separation at M-11 (132.5 Ma) from Antarctica (Powell et al., 1988). We identify the Mailaram High as the crest of the shoulder of the rift valley complex that became the initial northeast Indian Ocean (Fig. 44 in a later section). A hot spot subsequently may have generated the Rajmahal Trap, the Ninetyeast Ridge, and Kerguelen (Mahoney et al., 1983).

Late Cretaceous Stage

During this stage the Lameta Formation and the Infra-Trappean sediments (Fig. 29) were deposited during the relaxation (R9) that followed uplift (U8). The Lameta Formation and the Infra-Trappean sediments are interpreted as deposited on a semiarid alluvial plain (Brookfield and Sahni, 1987), contrary to the "marine" interpretation of Chanda (1968) and Robinson (1967). The upper part of the Lameta and the entire Infra-Trappean contain vertebrates that indicate a Maastrichtian age (Jaeger et al., 1989). From the single published paleocurrent vector (encircled 29 in Fig. 29), the Lameta Formation was deposited on a westward slope down the Narmada valley to overlap the locally eroded Jabalpur Formation to rest on Precambrian rocks. The other deposits of the Lameta and Infra-Trappean sediments are arranged in a simple drainage lobe in the Wardha area. Whether these outcrops formed part of a connected system or were isolated alluvial-lacustrine basins remains to be worked out.

In the Damodar area, faulting affected all rocks including the 120- to 110-Ma lamprophyre dykes (Sarkar et al., 1980) and is therefore dated as mid-Cretaceous or younger.

K/T (Cretaceous/Tertiary) Stage

The Deccan Trap was erupted about 65.5 Ma (= K/T boundary) over west and central India and on the east coast, as seen in a well at Narsapur (Jaeger et al., 1989, p. 317) (Fig. 30), near the early Paleocene shoreline. Outcrops of the basalt flows in the Narsapur area, known as the Rajahmundry Traps, have a $^{40}Ar/^{39}Ar$ age of 64.0 ± 0.4 Ma, marginally younger than the average age of 65.5 ± 0.5 Ma of the Deccan Trap but possibly coeval with lavas in the Kolhapur Formation of the Deccan

(text continues on p. 31)

Figure 16 (on page 24). Fence diagram showing time and space relations of the Gondwana basins, with the fence between Satpura (SAT) and Narsapur shown additionally out-of-perspective at the bottom. Data, including tectonic events (e.g., U5), are from Figures 4, 5, and 6. Timescale, with expanded scale from 200 to 300 Ma and shown in full on the middle part of the left-hand side, has these abbreviations: I to VI—palynologic compositions, AN—Anisian, B—Barakar, BM—Barren Measures, CA—Carnian, Carb—Carboniferous, E—Early, K—Karharbari, L—Late, LA—Ladinian, M—Middle, NEO—Neogene, NO—Norian, PA—Panchet, PAL—Paleogene, R—Raniganj, RH—Rhaetian, T—Talchir. Abbreviations of formations: B, BAR—Barakar, BI—Bijori, B/Y—Bhimaram/Yerrapalli, DE—Denwa, DH—Dharmaram, G—Gangapur, JAB—Jabalpur, K, KAR—Karharbari, M—Maleri, MO—Motur, PAC—Pachmarhi, PAR—Parsora, R—Raniganj, S-P—Supra-Panchet, TAL—Talchir; T-GO—Tirupati-Golapilli, ¢ indicates the carbonaceous shale of the Dubrajpur Formation. Abbreviations of place-names (indicated by dot in main part of diagram): A—Athgarh, AU—Auranga, BO—Bokaro, DA—Daltonganj, HU—Hutar, IB—Ib River, J—Jabalpur, JH—Jharia, KA—Kamptee, K-G—Krishna-Godavari, KOR—Korba, MAN—Manendragarh, MH—Mailaram High, NK—North Karanpura, P-G—Pranhita-Godavari (NW—northwest, SE—southeast), RA—Rajmahal, RAMK—Ramkola, RAN—Raniganj, SAT—Satpura, SI—Singrauli, TAL—Talcher, UM—Umaria, WV—Wardha Valley. Other: Cz-K—Cenozoic-Cretaceous.

Figure 17 (on page 25). Fence diagram showing variations in thickness of the Gondwana formations, with the fence between Satpura (SAT) and Krishna-Godavari (K-G) shown additionally out-of-perspective at the bottom. Abbreviations of place-names as in caption of Figure 16; symbols as on legend in Figure 16; formations are B, BAR—Barakar, BI—Bijori, BM—Barren Measures, DE—Denwa, DUB—Dubrajpur, JAB—Jabalpur, KAR—Karharbari, P—Panchet, PAC—Pachmarhi, R—Raniganj, SP—Supra-Panchet, T—Talchir. Other: P2—Paleocene.

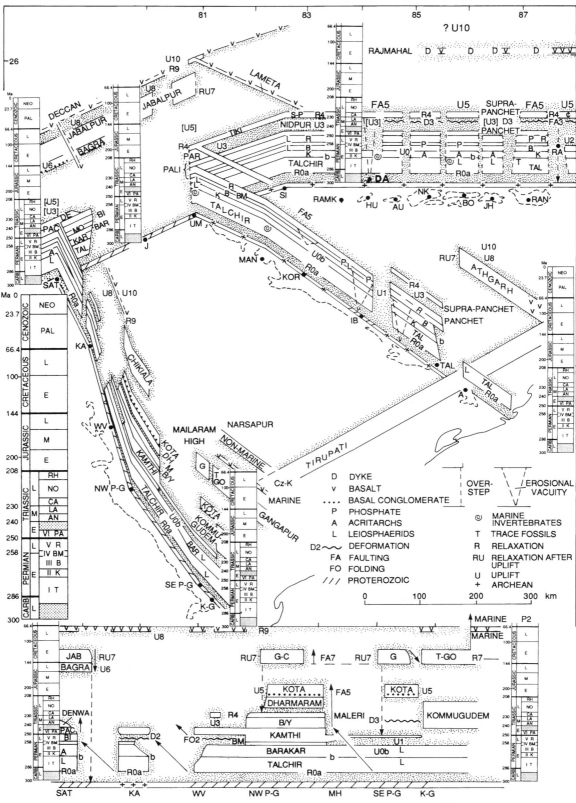

Figure 16. Caption on page 23.

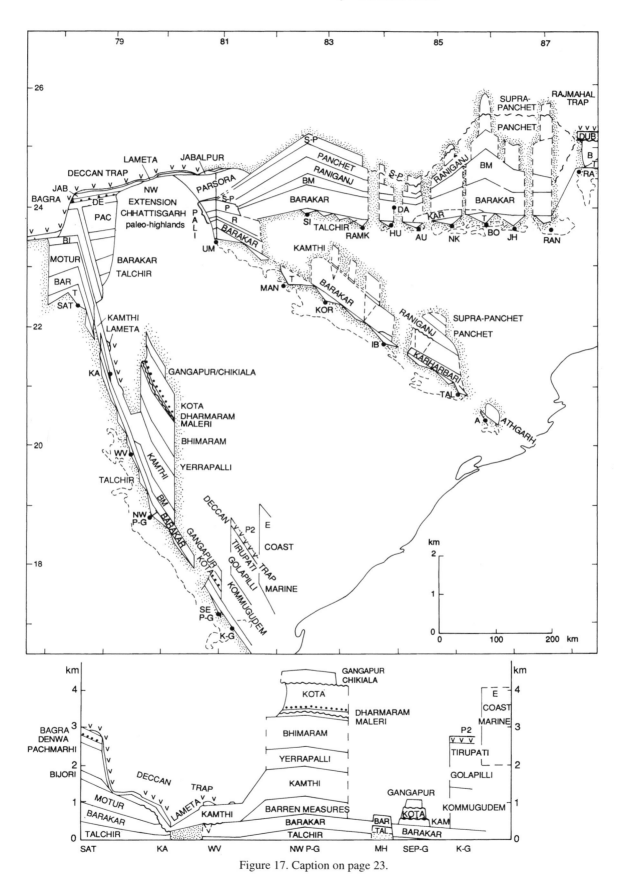

Figure 17. Caption on page 23.

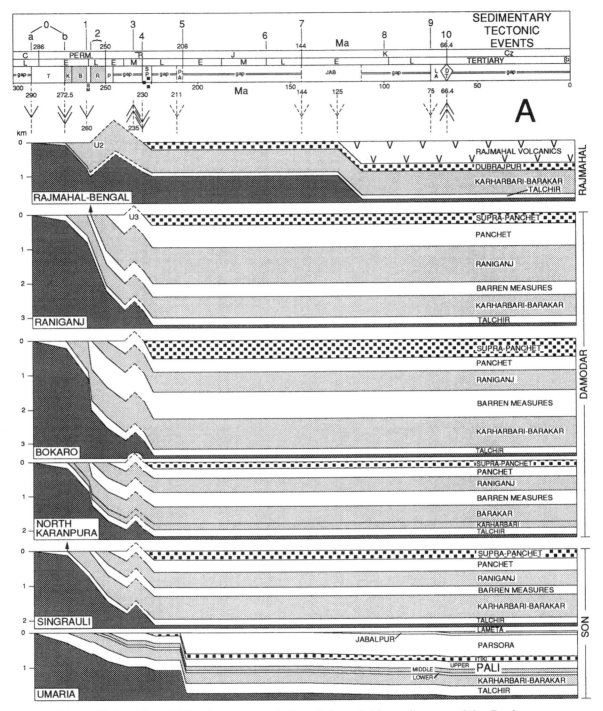

Figure 18 (on this and following two pages). Cumulative subsidence diagrams of the Gondwana basins of Peninsular India from 290 Ma to the present, with sedimentary-tectonic events 0 to 10 (Table 5) marked along the top. The 0-km line marks erosion or nondeposition above, deposition below. Uplift is sketched by broken lines, with the apex of the the triangle marking peak uplift. Abbreviations from the top left are C—Carboniferous, Perm.—Permian, R—Triassic, J—Jurassic, K—Cretaceous, Cz—Cenozoic; L—Late, E—Early, M—Middle, T—Tertiary, Q—Quaternary. Abbreviations of formations T—Talchir, K—Karharbari, B—Barakar, BM—Barren Measures, R—

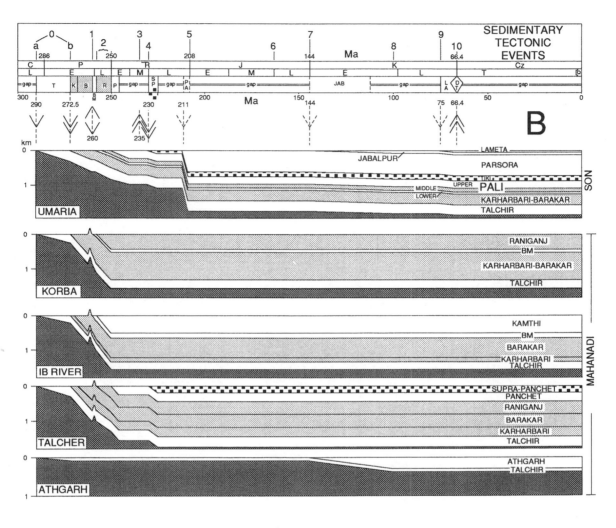

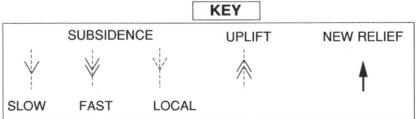

Raniganj, P—Panchet, SP—Supra-Panchet, PA—Parsora (taken here as latest Triassic), JAB—Jabalpur, LA—Lameta, and DT—Deccan Trap, and on Section C, B—Bhimaram, M—Maleri, and D—Dharmaram. Basement is filled with heavy stipple, coal measures with light stipple, Late Triassic rocks with a pattern of filled squares, and volcanics with Vs. Selected events (e.g., U3) are explained in Table 5. A, Rajmahal-Bengal basin above, Damodar basins in the middle, and Son basins below. B, Son basin above, Mahanadi basins below. C, Satpura, Kamptee, Wardha, northwest Pranhita-Godavari, and Godavari-Krishna basins. Abbreviations: PAL—Paleocene, EO-OLIGO—Eocene-Oligocene, MIO-PLIO—Miocene-Pliocene.

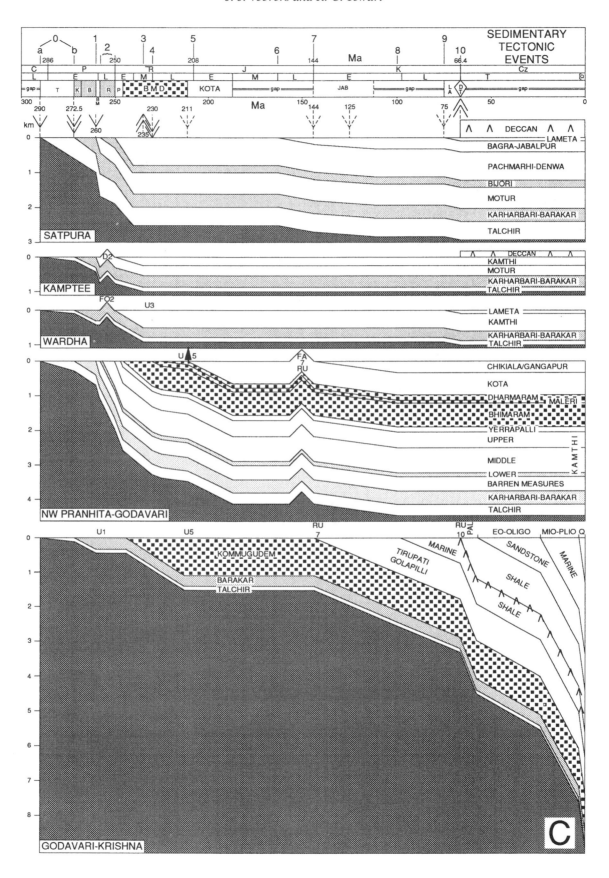

TABLE 5. TECTONIC AND SEDIMENTARY EVENTS

No.*	Type of Event	Age	Extent, Evidence
RU10	Uplift/stripping	Paleocene 66 Ma	Deccan, Rajmahal (clastic wedge offshore Mahanadi).
RI10	Rifting	Paleocene 66 Ma	Olpad fanglomerate Cambay Basin.
R9	Relaxation/subsidence	Latest K 75-66 Ma	Lameta Beds Satpura, Umaria-Bisrampur.
U8	Uplift?	Late K 100-70 Ma	Satpura, Umaria-Bisrampur
RU7 FA7	Relaxation/subsidence	Earliest K 144 Ma	Cambay, Satpura-Umaria, Pranhita-Godavari, East Coast (≈breakup eastern margin).
U7	Uplift	Earliest K 144 Ma	Mailaram High, Jabalpur
U6	Uplift	Late J 163 Ma	Satpura area north of Bagra Formation (= breakup western margin).
U5	Uplift	Earliest J 208 Ma	South of Kota Formation, P-G basin; Satpura and Umaria (?faulting); Koel-Damodar, Rajmahal.
FA5	Faulting	Earliest J 208 Ma	Right-lateral transcurrence Hutar, Raniganj; normal faulting Mailaram High between northwest and southeast Godavari.
R4	Relaxation/subsidence	Carnian 230 Ma	Supra-Panchet, Dubrajpur (=Pangean Extension II).
FA3	Faulting	Ladinian	Rajmahal, normal fault E (= Gondwanides II).
D3	Deformation	Ladinian	Southeast Godavari, Auranga-Bokaro, Raniganj.
U3	Uplift	Ladinian	Wardha, Umaria, Manendragarh, Talcher
[U3]	Uplift (inferred)	Ladinian	Satpura, Ramkola, Daltonganj, Jharia.
FO2	Folding	Raniganj	Wardha (= Gondwanides I).
D2	Deformation	258-250 Ma	Kamptee
U2	Uplift	258-250 Ma	Rajmahal, southeast Godavari.
U1	Uplift	Barren Measures 260-258 Ma	Korba-Ib-Talcher (southeast Mahanadi), southeast Godavari.
R1	Subsidence	260-258 Ma	All areas except those with U1.
U0	Uplift	Karh.-Barakar 272.5-260 Ma	Fans (post-glacial rebound): Tatapani, Auranga, N. Karanpura (Niyogi, 1987), Saharjuri, Hasdo-Arand, Mailaram (Mitra, 1987).
R0b	Rapid subsidence	Karh.-Barakar 272.5-260 Ma	All areas (acritarchs).
R0a	Relaxation/subsidence	Latest Carboniferous 290 Ma	Initial (Talchir) deposition, Gondwanas, 277-Ma glacioeustatic rise (= Pangean Extension I).

*D = deformation; FA = faulting, FO = folding, R = relaxation by subsidence; RI = rifting; RU = coupled relaxation and uplift; U = uplift.

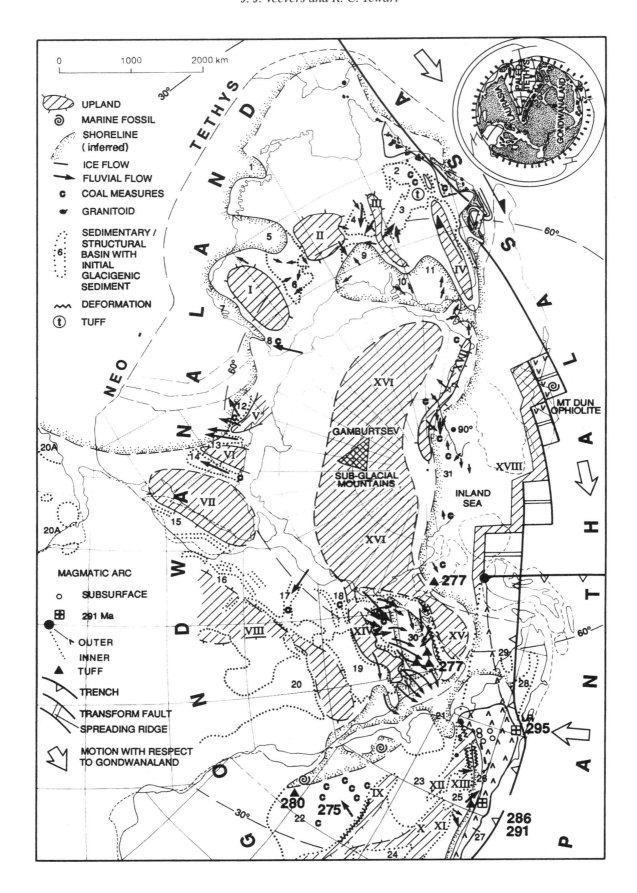

Figure 19. Paleotectonic/geographic map of the Gondwanaland province of Pangea during eastern Australian Stage A (290–268 Ma) centered about the Tastubian (275 Ma), including the 280-Ma interpolated pole and paleolatitudes, from Veevers et al. (1994c). References to the South Pole and latitudes correspond to present coordinates. The global setting (top right corner, from de Wit et al., 1988), was such that Gondwanaland, the southern province of Pangea, lay between Panthalassa on one side and the equatorial Paleotethys and Laurasia on the other. Shortly after the 320-Ma amalgamation of Pangea, the Cimmerian continent (C) was transferred from the Gondwanaland margin to Laurasia by the generation of Neotethys and the consumption of Paleotethys (Veevers, 1988a). A key to paleogeographic features is given in Table 6.

Trap near the west coast (Baksi et al., 1994). Subsequent erosion has removed trap from the Satpura Range and along the Son basin and the northern part of the Mahanadi basin, leaving the big northwest Chhattisgarh area of trap in between. The trap remains effectively horizontal, unaffected by subsequent movement except on the west coast, where rifting (R10) in the Cambay Basin has deeply depressed the trap beneath the Olpad fanglomerate and later deposits (Ramanathan, 1981). Baksi et al. (1994) suggest that the Rajahmundry Traps represent remnants of intracanyon flows eastward down the ancestral river systems of Peninsular India.

TABLE 6. KEY TO FIGURE 19*

	Australia					
I	Ancestral Great Western Plateau	14	*Godavari*	26	*San Rafael*
II	*Central Australia*			27	Calingasta-Uspallata
III	*Ancestral South Australian Highlands*		**East Africa**	28	*Tepuel*
IV	*Ancestral eastern Australian foreswell*	15	*Morondava*	29	*Golondrina*
1	Bonaparte Gulf	16	*Tanzania Karoo*	IX	*Asuncion*
2	*Galilee*	VII	Madagascar-?southwest India	X	*Michicola*
3	*Cooper*			XI	*Puna*
4	*Pedirka*		**Central Africa**	XII	*Pampean*
5	Canning	VIII	Congo-Kaokoveld core of ice sheet on ?upland.	XIII	*Pie de Palo*
6	Officer	17	*Lower Zambezi*		**Southern Africa**
7	Carnarvon-*Perth*	18	*Waterberg*	30	*Karoo*
8	*Collie*	19	*Kalahari-Botswana*	XIV	*Cargonian*
9	*Arckaringa*	20	*Congo*	XV	*Proto-Fold Belt*
10	*Troubridge*	20A	*Arabia*		**Antarctica**
11	*Oaklands, Renmark*		**South America**	31	Transantarctic
	India	21	*Sauce Grande*	XVI	*East Antarctic*, centered on the Gamburtsev Subglacial Mountains
V	Chotanagpur	22	*Paraná*		
VI	Chhattisgarh	23	*Chaco-Paraná*	XVII	Ross High
12	*Damodar*	24	*Tarija*	XVIII	Postulated Antarctic margin
13	*Mahanadi*	25	*Paganzo*		

*Shown in the platform area outside the Panthalassan margin are new *(in italics)* or renewed basins (e.g., 1) and uplands (e.g., III).

TABLE 7. SOURCE OF DATA PLOTTED ON FIGURES 20 THROUGH 30

1	Smith, 1963a, b.	21	Srinivasa-Rao et al., 1979.
2	Frakes et al., 1975.	22	Ramanamurthy, 1985.
3	Casshyap, 1979a, citing Srivastava, 1970.	23	Reddy and Prasad, 1988.
4	Casshyap, 1979a, citing Ahmad, 1975; Ahmad and Hashimi, 1974, 1976; Ahmad et al., 1976; also Ahmad, 1981.	24	Casshyap, 1981b.
		25	Casshyap and Kumar, 1987.
		26	Casshyap and Tewari, 1988.
5	Casshyap and Qidwai, 1974.	27	Sen and Sinha, 1985. (i) We regard the crossbedding as aqueous; (ii) we group the vector means for the upper part of the Supra-Panchet (sectors 2-4) to determine a grand mean of north-northeast.
6	Casshyap and Tewari, 1982.		
7	Casshyap and Srivastava, 1988.		
8	Datta et al., 1979.		
9	Bose and Ramanamurthy, 1979.		
10	Ghosh and Mitra, 1970b.		
11	Sengupta, 1970.	28	Rudra, 1982.
12	Casshyap and Qidwai, 1971.	29	Chanda, 1968.
13	Casshyap, 1973.	30	Prasada-Rao, 1976.
14	Casshyap, 1979b.	31	Kumar and Bhandari, 1973.
15	Casshyap and Tewari, 1984.	32	Ranga-Rao et al., 1979, p. 485.
16	Tewari and Casshyap, 1982.	33	Casshyap et al., 1993a.
17	Khan, 1987.	34	Chaudhuri and Mondal, 1989.
18	Khan and Casshyap, 1979.	35	Dutta and Laha, 1979.
19	Khan and Casshyap, 1982.	36	Tewari, 1992.
20	Rishi, 1972.	37	Tewari and Casshyap, 1994.
		38	Casshyap et al., 1993a.

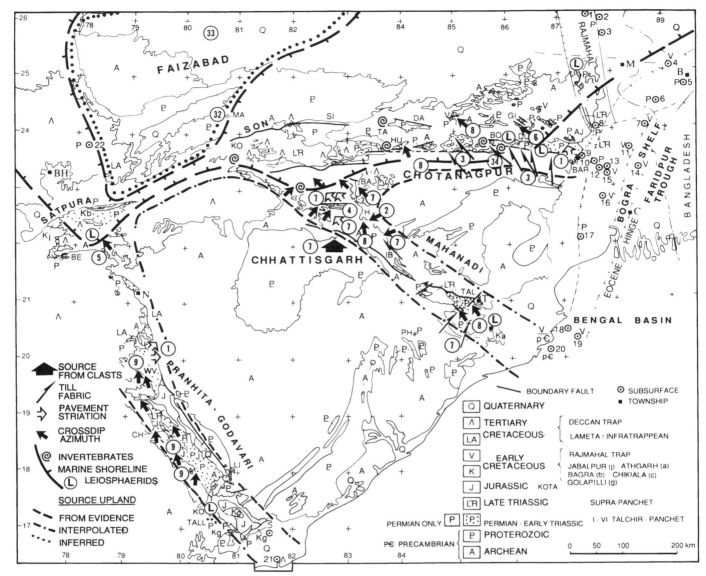

Figure 20. Paleogeography during the deposition of the Talchir Formation plotted on the base map of Figure 2. The encircled numerals denote the source of information (Table 7). The subsurface occurrence (22) on the western side of the Faizabad upland is at Mirkheri (Fig. 2); another isolated occurrence, on the eastern side of the Faizabad upland, is at Maihur (MA).

In the final event mentioned here, the Deccan plateau was uplifted (U10) some 2,000 m (Radhakrishna, 1989) and subsequently worn back from its original extent, with sediment from the uplift deposited in a clastic wedge in the offshore Mahanadi basin.

CONNECTIONS WITH TETHYS AND GONDWANALAND

Global scene at the 250-Ma Permian/Triassic boundary

In Pangea, the Gondwana master basin occupied the midlatitude ground between the Tethyan margin and the adjacent Western Australian, East Antarctic, and Arabia–Madagascar–East African parts of Gondwanaland (Fig. 31).

The connections are shown in more detail in the following Figures 32 through 39, which show the regional paleogeography of Peninsular India and adjacent parts of Tethys and Gondwanaland from the mid-Carboniferous start of Pangea (Fig. 32) to its breakup at the Jurassic/Cretaceous boundary (Fig. 39). The base map and paleolatitude of Figures 32 to 38 are from Powell and Li (1994), and Figure 39 is from de Wit et al. (1988) and Veevers et al. (1991). Details of Peninsular India are from previous chapters; those of adjacent areas are from Veevers and Powell (1987), augmented by references cited in the text.

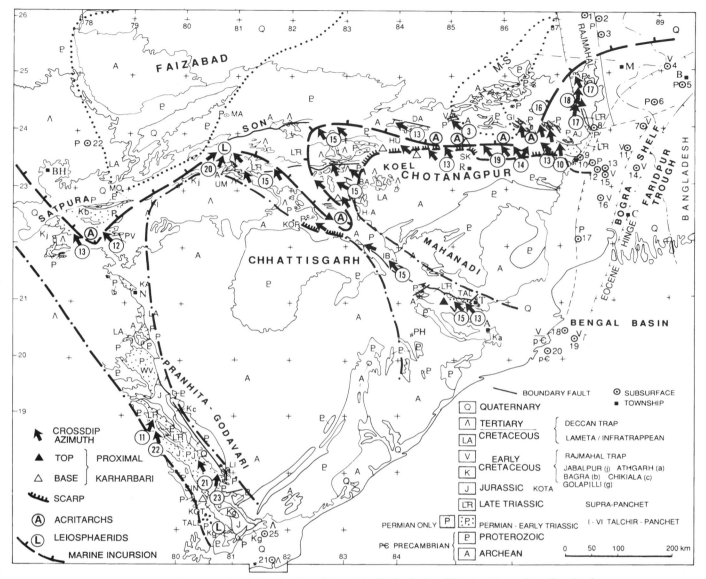

Figure 21. Paleogeography during the deposition of the Karharbari and Barakar Formations. Proximal facies of the Karharbari Formation: base, shown by an open triangle, from Gee (1932), Mitra (1987), and Niyogi (1987); top, shown by a filled triangle, from De (1979). The encircled numerals denote the source of information (Table 7). Upland symbols as in Figure 20. M-S—Monghyr-Saharsa ridge.

Late Carboniferous (320 to 290 Ma)

The Late Carboniferous epoch (Fig. 32) followed the Variscan collision of Gondwanaland and Laurussia (Veevers et al., 1994d) and the far-field inversion by crustal shortening of the Amadeus Transverse Zone (Alice Springs Orogeny) of Central Australia and, speculatively, of East Antarctica, represented today by the Gamburtsev Subglacial Mountains (I) (Veevers, 1994b). The sea was confined to the immediate Paleotethys margin except in northern Australia. The rest of the land area is represented by a lacuna in the stratigraphic record, probably because it was above base level during the initial accumulation of Pangean self-induced heat (Veevers, 1994a). Sea level had dropped from its Devonian–early Carboniferous high, possibly by glacio-eustatic drawdown, promoted by the buildup of ice on the polar plateau (I) and on other high ground in South America and in central and eastern Australia (Veevers and Powell, 1987).

Carboniferous/Permian boundary (290 to 275 Ma)

What we call the Indo-Australian rift zone (Fig. 33A) lay behind the Afghan, Lhasa, and Sibumasu blocks that drifted northward (present coordinates) at the end of the Permian. The only part of this zone that has remained intact and accessible is

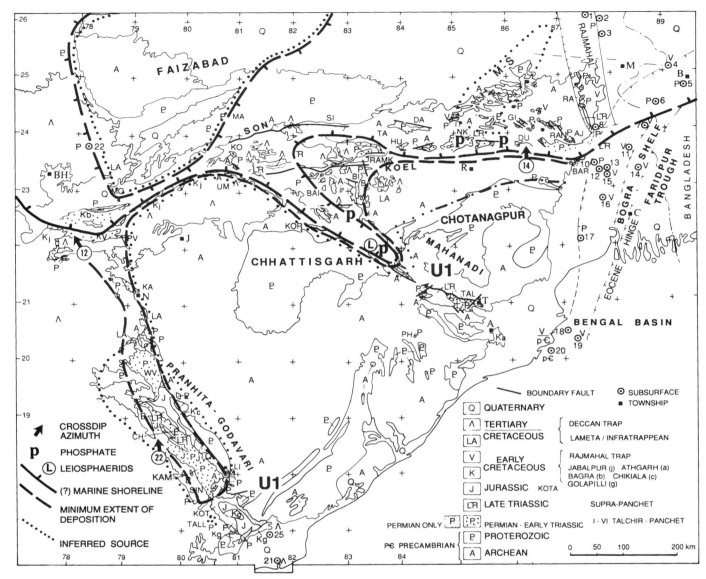

Figure 22. Paleogeography during the deposition of the Barren Measures. The encircled numerals denote the source of information (Table 7). U1 is explained in Table 5. M-S—Monghyr-Saharsa ridge.

the passive margin of Western Australia. According to Etheridge and O'Brien (1994), initial rifting was in Late Carboniferous–Early Permian time, involved 100% to 400% northwest-southeast extension beneath most of the present continental shelf of Western Australia, and was accompanied by a thick Permian and Triassic "sag" sequence that makes up more than 50% of the total sediment column. The western side of the Indo-Australian rift zone was bordered by the Madagascar rift zone, which lay behind the narrow Permotethys (Stampfli et al., 1991). Wopfner (1993) explained the distribution of the graben complex of East Africa and Madagascar by a transtensional stress-field (shown by the arrows in Fig. 36) operating in Permo-Carboniferous and Early Triassic times. The left-lateral components of this stress (small double-shafted arrows in Fig. 36) are consistent with left-lateral movement generated by the opening of Permotethys along a continent-continent transform fault between the Madagascar rift zone and India (large double-shafted arrows). All but the top right and lower left parts of the map area lie within the zone (between the dotted lines) of Zambezian-type or faulted basins, flanked by the Karoo-type or flexed basins (Rust, 1975; Tewari and Veevers, 1993, fig. 2).

Widespread glaciation ensued during the rest of the Late Carboniferous and pre-Sakmarian Permian. The ancestral Gamburtsev Mountains (I) formed the focus of radial ice drainage into Australia, India, and Africa (Tewari and Veevers, 1993; Veevers, 1994b). The ice flowed through the uplands of the Pilbara-Yilgarn block in Western Australia (II) and the central Australian upland (just east of the map area), the Chotanagpur (III),

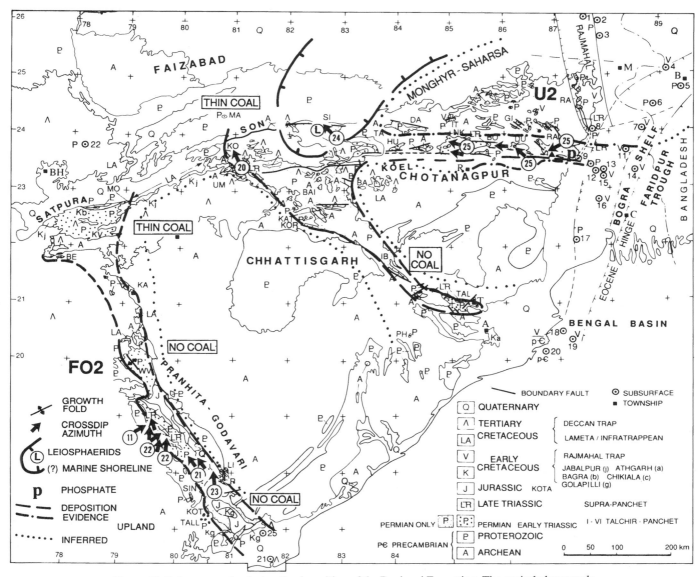

Figure 23. Paleogeography during the deposition of the Raniganj Formation. The encircled numerals denote the source of information (Table 7). U2 and FO2 are explained in Table 5.

Chhattisgarh (IV), Faizabad (V) and south Indian-Madagascar (VI), and the Congo-Kaokofeld (VII) and Cargonian (VIII) uplands of Africa. Negligible glacigenic sediment remained on the continental platform until the first release of Pangean heat at 290 Ma (base of East Australian palynologic Stage 2 or Gzelian) generated accommodation space by the initial subsidence of the Gondwana basins (Veevers, 1994a). Retreat of the ice at 275 Ma (Tastubian or early Sakmarian *Eurydesma* zone; Dickins, 1984) brought the sea into Gondwanaland. From the Arabian margin (Murris, 1980) of Permotethys and from the Afghan-Lhasa-Sibumasu margin of Paleotethys, the sea transgressed across the Indo-Australian rift zone into central India to southwestern Australia and around upland II into the Canning Basin. On the other side of Gondwanaland, a marine transgression entered the Karoo basin of southern Africa from Panthalassa. In central India and southern Africa, the short-lived Tastubian transgression was succeeded by a nonmarine succession of Permian coal measures and Triassic barren strata, whereas in the Indo-Australian rift zone the sea remained, at least intermittently, during the rest of Permian and Triassic time, as shown by the Westralian basin (Bradshaw et al., 1988). We explain this difference by suggesting that in central India and southern Africa the rising sea covered the formerly glaciated ground until isostatic rebound caused it to regress. The extension of the crust in East Africa–Madagascar, India, and Western Australia led to the subsidence of many areas below sea level. As shown in Figure 33B, the occurrence of volcanics above the glacigenic deposit in the Agglomeratic Slate of Kashmir and in the Woniusi Formation of

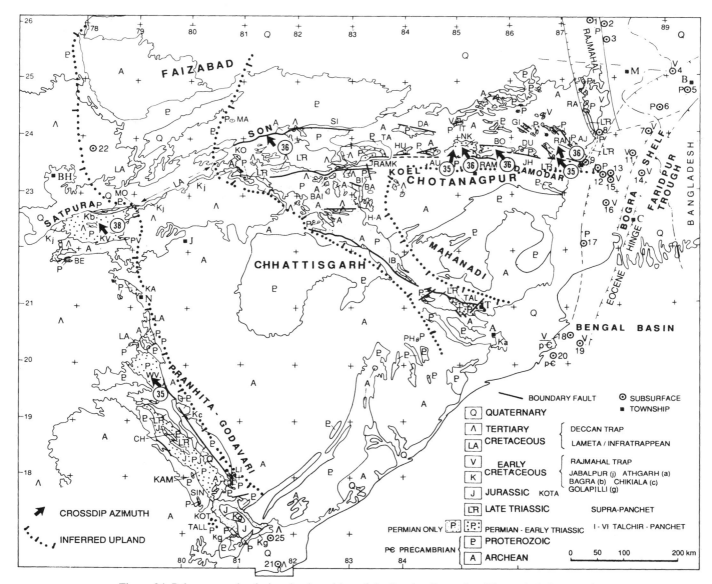

Figure 24. Paleogeography during the deposition of the Panchet Formation. The encircled numerals denote the source of information (Table 7). Not shown is the information on the Mesozoic vectors, which is given by Tewari and Casshyap (1994) (#37).

western Yunnan and the emplacement of the anorogenic Yuman, Malakand, and Ambela granites (Spring et al., 1993) match the succession in eastern Australia (Veevers et al., 1994e), and both sets of volcanics and granitoids match those of Europe (Veevers et al., 1994f), all characteristic of the initial extension of Pangea (Veevers and Tewari, 1995).

Artinskian (265 Ma)

Marine invertebrates in Arabia, in the Badhaura Formation at Bap, in the Amb Formation of the Salt Range, in the Garu Formation in Siang and at Namchi (Sikkim), and in Western Australia (Fig. 34) indicate an open sea, delimited by the barbed line around the Faizabad upland (V); the barbed broken line delimits the coastal, brackish environment of the northern Gondwana coal basin. Drainage in the Damodar area was deflected around the newly risen Monghyr-Saharsa upland (II). The broad open arrow in the Amery area of East Antarctica signifies the through flow of sediment above base level from the Gamburtsev upland (I). The Ecca coal measures in southern Africa prograded from the Witwatersrand Arch (VIII) and from the east.

Kungurian-Ufimian (260 Ma)

A narrow arm of the Tethyan sea penetrated the Madagascar rift zone as far as southern Madagascar; a broad embayment occupied the Indo-Australian rift zone, at the coastal head of which the Barren Measures were deposited in the northern part

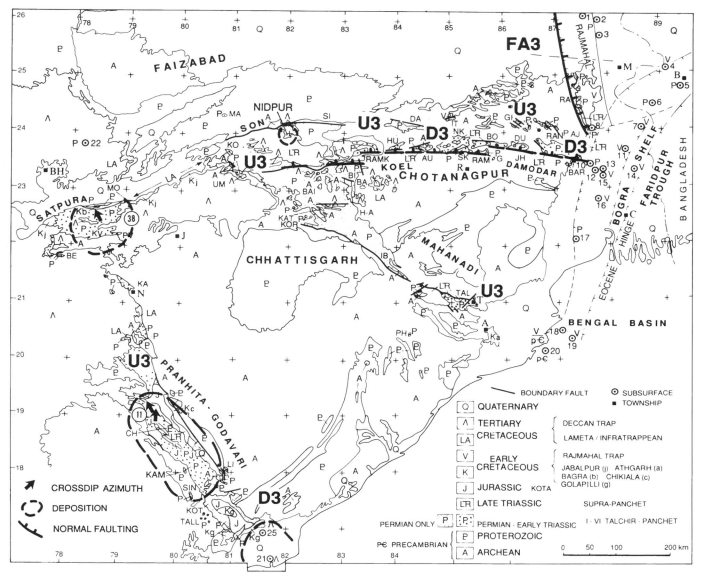

Figure 25. Paleogeography at the end of the mid-Triassic (Anisian-Ladinian), except the paleocurrent vector for the slightly earlier (late Scythian) Denwa Formation (#38). The encircled numerals denote the source of information (Table 7). The Triassic age of sediments in the Krishna-Godavari area is from Prasad and Jain (1994). D3, FA3, and U3 are explained in Table 5.

of the Gondwana master basin; and the sea covered part of Western Australia (Fig. 35). In the Amery area of East Antarctica, deposition started with the Radok Conglomerate (R) shed from the sides of a graben that came off the Gamburtsev upland (I). The upper part of the Ecca Group of the Karoo basin prograded northeastward from the rising Cape foldbelt. Deposition continued in East Africa and Madagascar but stopped in the Congo basin.

Tatarian (253 Ma)

At Beaver Lake (BL in Fig. 36), the braided-river deposit of the Bainmedart Coal Measures and the lower part of the Flagstone Bench Formation, both with east Australian Stage 5 palynomorphs, contain a northerly crossdip azimuth (Webb and Fielding, 1993). Downstream from Beaver Lake, the Saharsa-Monghyr upland (II) rose across the northwest drainage, which became confined to a west-sloping valley that possibly emptied in an estuary. To the west, the long gulf from Permotethys persisted. Corals in the Vohipanana-Ambatokapika limestone in the Lower Sakamena Group of southern Madagascar (Boast and Nairn, 1982) are dated by palynomorphs (Wright and Askin, 1987) as equivalent to the Late Permian Lower Beaufort Group of southern Africa; and marine strata with *Cyclolobus* and *Productus* in northern Madagascar, the Ruhende Beds of

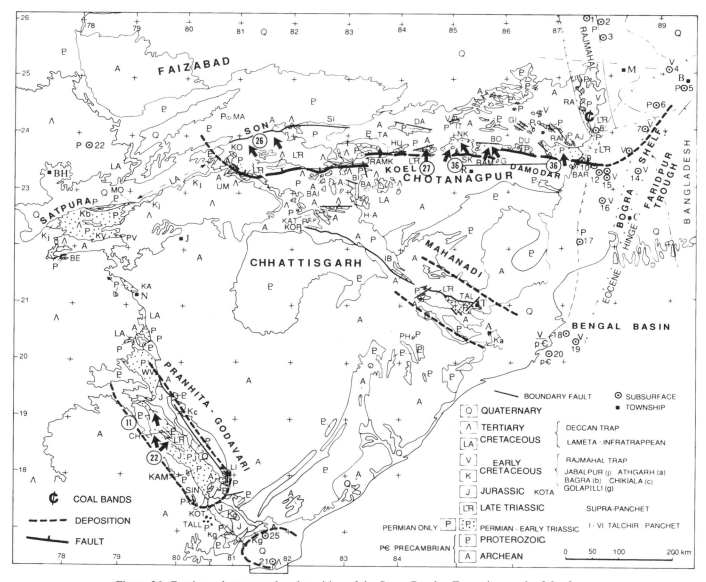

Figure 26. Carnian paleogeography: deposition of the Supra-Panchet Formation north of the downfaulted Son and Koel-Damodar areas and age-equivalents in the Mahanadi and Godavari areas. The encircled numerals denote the source of information (Table 7).

the Mikumi Basin (Kreuser et al., 1990), and the marine horizon in the Nyakatitu Basin (Yemane and Kelts, 1990) in Tanzania are equivalent to the marine Chhidru Formation of the Salt Range (Veevers et al., 1994e).

Widespread volcanism behind Tethys included late Murgabian (255 to 253 Ma) basaltic trap and other intraplate volcanics erupted during rifting in the shelf succession of Oman (Robertson and Searle, 1990; Rabu et al., 1990), the Murgabian-Midian (Kazanian–early Tatarian, 255 to 251 Ma) Panjal Trap immediately south of the Lhasa block in Zanskar (Stampfli et al., 1991), and rhyolite, dolerite, and undersaturated rocks in Western Australia, all anticipating breakup in the north (Fig. 33B).

In Western Australia, nosean phonolite, lamprophyre, and trachyte were emplaced in Edel-1 in the northern Perth basin and, probably at this time, rhyolite in Enderby-1 in the Dampier basin (Geological Survey of Western Australia, 1990, p. 595). The volcanics in Edel-1 are regarded as flows and dikes of undersaturated alkaline rocks similar geochemically to volcanics along the Atlantic rifted margin (Le Maitre, 1975). The volcanics lie on either side of an angular unconformity called WA-U3 (Smith and Cowley, 1987; Bergmark and Evans, 1987), which separates the Late Permian (mid-Kazanian, 255 Ma) Wagina Sandstone and older strata affected by north-striking, east-dipping extensional faults from the "basal Triassic sandstone" and overlying Kockatea Shale (>245 Ma) and younger strata affected by northwest-southeast, high-angle, west-

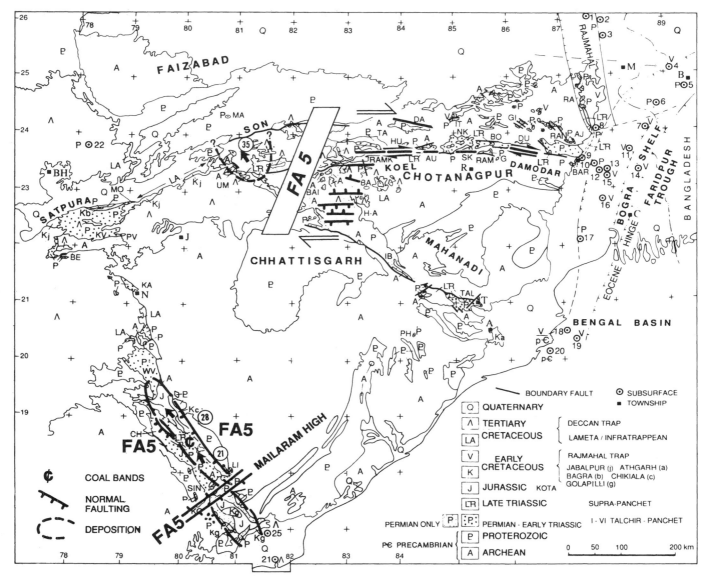

Figure 27. Early Jurassic paleogeography: deposition of the Kota Formation in the Pranhita-Godavari-Krishna area (Murti and Lakshminarayana, 1994) and right-lateral transpressional faulting in the Son-Mahanadi and Koel-Damodar region. The encircled numerals denote the source of information (Table 7). FA5 is explained in Table 5.

dipping planar normal faults (Smith and Cowley, 1987, fig. 4). This break is general along the western and northwestern Australian margin, from the Perth basin in the south to the Bonaparte basin in the north (Boote and Kirk, 1989, fig. 2). We interpret WA-U3 as a breakup unconformity that marks the end of the locally intense phase of rift extension and the start of breakup of the Paleotethyan margin (Afghan, Lhasa, Sibumasu blocks) by the generation of Neotethys (Fig. 37). K-Ar dates of 261 ± 5 Ma and 267 ± 5 Ma from samples below the unconformity (Smith and Cowley, 1987, p. 122, 123) suggest ages from Sakmarian to Kazanian, and we adopt the younger limit appropriate to the probably coeval Kazanian Wagina Sandstone. In the adjacent Wittecarra-1 (W-1) well, lamprophyric flows or shallow sills in the Early Triassic Kockatea Formation indicate Triassic volcanism, so that those volcanics above the unconformity in Edel-1 may be Triassic. The section of rhyolite and sediment in Enderby-1 extends from a depth of 2,083 m to the total depth of 2,150 m. The associated (?interlayered) sediment contains marine microplankton similar to that in the earliest Triassic basal Locker Shale (Tony Bint, personal communication, 1994). The rhyolite could extend below the total depth and hence into the Permian. In the Canning basin, doleritic rocks (Reeckmann and Mebberson, 1984) are widely distributed offshore and in the Broome and Derby areas onshore (Fig. 36).

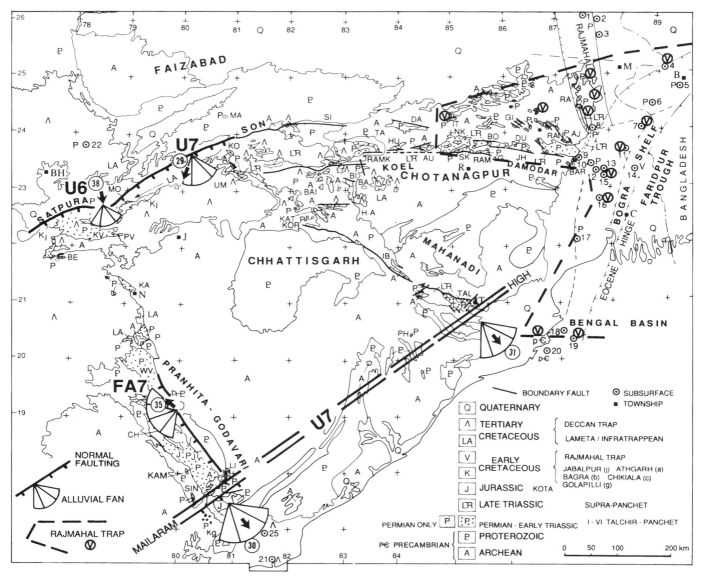

Figure 28. Paleogeography during Late Jurassic (Bagra Formation) and Early Cretaceous times. The encircled numerals denote the source of information (Table 7). U6, U7, and FA7 are explained in Table 5.

Gleadow and Duddy (1984) applied the fission-track method to dating apatite in the intruded rock and in the dolerite and found a mean minimum age of 255 ± 25 Ma or 280 to 230 Ma. The dolerite is younger than the youngest intruded rock, the 263-Ma Noonkanbah Formation, so its age falls between 263 and 230 Ma. In Perindi-1, apatites derived ultimately from the basement (with abundant uranium) gave an annealing date of 251 ± 15 Ma; and a dolerite in Pearl-1 gave a K-Ar date of 249 ± 2 Ma, all consistent with an age of 255 to 250 Ma, near the end of the Permian. The Permo-Triassic magmas along the divergent plate boundary of Neotethys together with those from the convergent plate boundary of Panthalassa to the Siberian Traps (Fig. 33B) mark a conjunction of diverse and exceptionally voluminous magmatism that probably triggered the environmental catastrophe at the Permian-Triassic boundary (Veevers and Tewari, 1995).

Another low-angle Permian-Triassic unconformity is seen in East Africa (Wopfner, 1993). The unconformity separates the Late Permian K6 and the Early Triassic Kingori Sandstone (K7) (Fig. 37) and marks a change from normal faults to listric, extensional faults, the reverse order of style of faulting in the Perth Basin. Wopfner (1993) explained the distribution of the graben complex of East Africa and Madagascar by a transtensional stress-field (shown by the arrows in Fig. 36) operating in Permo-Carboniferous and Early Triassic times. The left-lateral components of this stress (small double-shafted arrows in Fig. 36) are

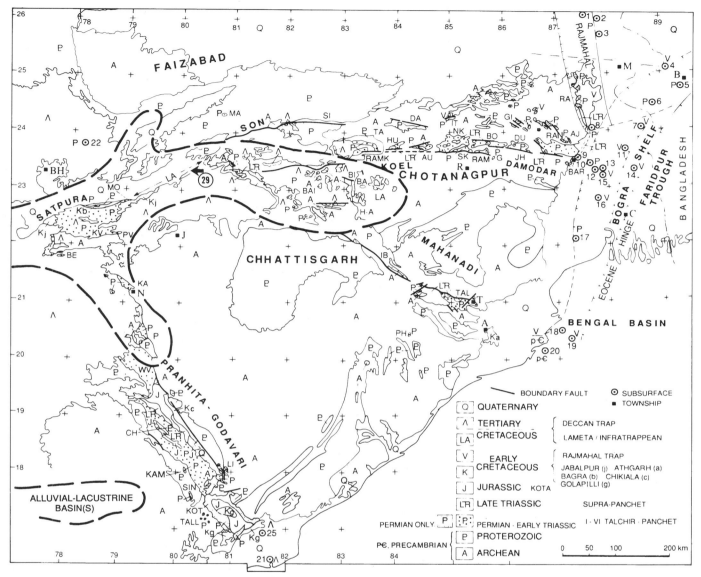

Figure 29. Paleogeography during Late Cretaceous time. The encircled numeral denotes the source of information (Table 7).

consistent with left-lateral movement generated by the opening of Permotethys along a continent-continent transform fault between the Madagascar rift zone and India (large double-shafted arrows). A marine intercalation in the Maji-ja-Chumvi Formation of Kenya (Kreuser et al., 1990) is correlated with northern Madagascar. Yemane and Kelts (1990) suggest that much of central and southern Africa was covered by giant freshwater lakes, which contained the nonmarine bivalve *Kidodia coxi*.

In southern Africa, alluvial lobes of the lower Beaufort Group were shed from rapidly rising uplands in the east and south (Outeniqua folding event of the Cape Fold Belt). At the same time, in the Tatarian (253 to 250 Ma), ash-flow tuff and pumice lapilli (v in Fig. 36) were erupted in the Heilbron-Frankfort area of the Orange Free State (Keyser and Zawada, 1988), during proximal pyroclastic volcanism, unrelated to the distal volcanism from the Panthalassan magmatic arc (Veevers et al., 1994e).

Other basalt eruptions at this time (Fig. 33B) were the extensive 100-m-thick Longtanian (252 to 251 Ma) Omeishan Basalt of the southwest Guizhou region of South China (Zunyi et al., 1986, p. 117) and the vast end-Permian (250 Ma) Siberian Traps (Campbell et al., 1992).

Early Scythian (247 Ma)

Deposition of the Jetty Member, the middle part of the Flagstone Bench Formation with redbeds at Beaver Lake in East Antarctica, probably persisted into the Early Triassic (Fig. 37), though without diagnostic fossils its age is known no more

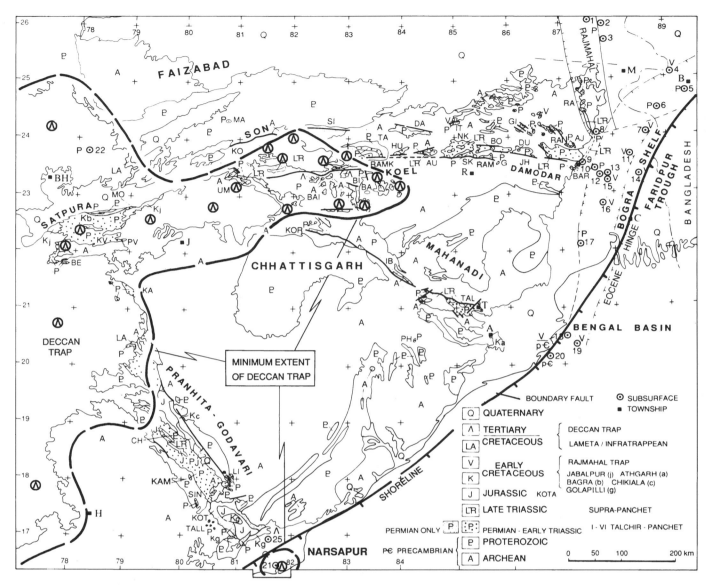

Figure 30. Paleogeography about the Cretaceous (K)/Tertiary (T) boundary. Shoreline from data in Alam et al. (1990), Jagannathan et al. (1983), and Jaeger et al. (1989).

closely than its being bracketed by Stage 5 (latest Permian) lower Flagstone Bench Formation below and Norian upper Flagstone Bench Formation above (Foster et al., 1994). Redbeds in Ocean Drilling Program drill-hole 740 in Prydz Bay, probably equivalent to the middle part of the Flagstone Bench Formation (Turner, 1991; Keating and Sakai, 1991), were derived from an uplift southeast of the drill site (Turner, 1991) by renewed uplift of the ancestral Gamburtsev Mountains (I). Downstream in India, the Gondwana master basin was disrupted by the renewed uplift of the Chotanagpur upland (II) and by the coalescence of the Chhattisgarh and Faizabad uplands (III). The paleoslope was wholly to the northwest and north, and the earlier Permian estuarine arm of the Tethyan sea is no longer evident. However, the marine gulf to the west persisted, as indicated by the *Claraia* shales of northern Madagascar and by the marine Maji-ya-Chumvi Formation of Kenya (Veevers et al., 1994e). The low-angle unconformity in East Africa between K6 and K7 is interpreted by Wopfner (1993) as indicating an easterly tilt by detachment of the eastern rift margin. Alluvial lobes of the upper Beaufort Group were shed from the even more rapidly rising uplands that followed the 247-Ma folding and thrusting of the Ecca and lower Beaufort rocks of the Cape Fold Belt.

In Western Australia, the undersaturated volcanism in Edel-1 was joined by the eruption (as a flow or sill) of lamprophyre in Wittecarra-1 (Smith and Cowley, 1987) and by the

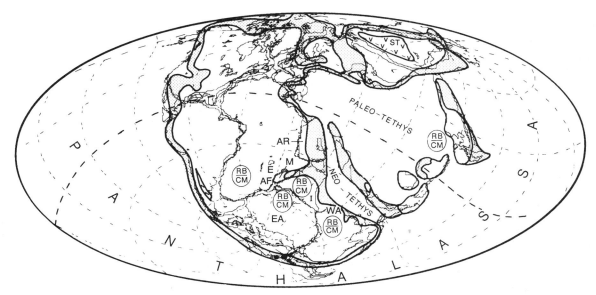

Figure 31. Global scene at the Permian/Triassic boundary, 250 Ma. From Veevers (1994a). RB/CM indicates those areas, including Peninsular India, in which redbeds (RB) replaced coal measures (CM). Broken heavy line denotes equator. ST—Siberian Traps (V pattern). Shading indicates shallow seas. AR—Arabia, EA—East Antarctica, EAf—East Africa, I—India, M—Madagascar, WA—Western Australia.

rhyolite in Enderby-1 above the regional unconformity called WA-U3. Alkaline basaltic flows and pyroclastics in Ladakh (v) (Honegger et al., 1982) and the Gaik granite (filled circle) (Spring et al., 1993), restored from their present (boxed) position to the prethrusted position, were erupted along the margin during the opening of Neotethys (Fig. 33B).

Late Triassic (230 to 208 Ma)

Figure 38 shows the depositional events (R4 of Table 5) that followed deformation and uplift (D3, U3, of Table 5) in the Ladinian and finally the right-lateral transcurrence (FA5) in the earliest Jurassic.

A dinocyst and the Norian freshwater alga *Bartenia communis* (Bc) (Foster et al., 1994) together with the *Dicroidium* flora (DI) (Webb and Fielding, 1993) occur in the upper Flagstone Bench Formation at Beaver Lake. The dinocyst is unknown elsewhere, but *Bartenia communis* is known in the Rankin area (R) of the northern Carnarvon basin as an element of the swamp palynofacies that occurs in close proximity to samples yielding dinoflagellates, suggesting a lower-delta-plain environment behind a complex marine shoreline (Bint and Helby, 1988). Accordingly, a possible connection to Neotethys in the Norian is indicated in Figure 38 by a narrow shallow gulf between Western Australia–Antarctica and Peninsular India. The uplift (U in Fig. 38) west of Beaver Lake is inferred from the eastward crossdip azimuth in the middle part (Jetty Member) of the Flagstone Bench Member, which was deposited in side streams (Webb and Fielding, 1993). In the overlying upper Flagstone Bench Formation, the azimuth reverts to the north, interpreted as the flow of the trunk stream.

Downslope in India, the Carnian Supra-Panchet Stage was deposited during relaxation/subsidence (R4) on northerly paleoslopes after faulting (FA3), deformation (D3), and uplift (U3) of the Chotanagpur upland (II) further dismembered the master basin.

In Western Australia, late Ladinian (233 Ma) right-lateral transpression (FA3) in the northern Canning Basin generated fold uplifts that were cut by rift faults in the Carnian (230 Ma). Sediment from the uplifted Pilbara Block (PB) formed the Rankin delta (R) that was impounded by a downslope uplift, and to the south the Leseuer Sandstone (LS) was deposited in the graben of the Perth Basin (Veevers, 1988a, fig. 4C). Rift volcanics were emplaced along the uplifted margin of the Exmouth Plateau (EP), including the Mount Victoria Land block (MVL), and Rhaetian (Rh) (210 Ma) carbonate reefs developed along the northern margin (Exon et al., 1992; von Rad et al., 1992a, b). The Onslow Palynoflora (OP) of the Carnarvon Basin is found also in the Krishna-Godavari basin and in northern Madagascar (Prasad and Jain, 1994).

In Africa, after a Middle Triassic lacuna, the 6-km-thick nonmarine sandstone of the Isalo Group of Madagascar was deposited on a low-angle unconformity of faulted Zambezian terrain in the downwarp between Madagascar and East Africa and thinned to the west. Likewise, alluvial lobes of the Molteno Formation were shed from the Cape Fold Belt, which was terminally folded in the Ladinian Gondwanides II (Veevers et al., 1994c). Other alluvial lobes were shed from uplands in the east and north (Visser, 1984).

Deposition of coal in Pangea resumed in the Carnian after the Early and Middle coal gap (Veevers et al., 1994a) in areas that subsided within the Gondwanides or on the platform. Coal measures were deposited in Australia at Ipswich and Leigh Creek and in Tasmania and in Africa in the Molteno Formation; in India, carbonaceous matter, not seen since the Permian, was deposited and preserved in the Dubrajpur Formation.

The compressional deformation at 233 Ma on the Gondwanaland platform, including the U3 event in India (Table 5), corresponds to the Gondwanides II event along the Panthalassan margin of Gondwanaland (Veevers et al., 1994c); and the 230-Ma subsidence by Pangea-wide rifting (Veevers, 1994a) corresponds to event R4 in India (Table 5).

Earliest Jurassic deformation

The coarse grade of the fluvial sandstone of the Early Jurassic Kota Formation in the Godavari basin is interpreted as indicating intermittent uplift of the source area (Rudra, 1982, p. 77). Mitra (1987, p. 36) interprets the basal pebbly sandstone as a debris-flow deposit that indicates activation of the northeast border fault in the Early Jurassic (FA5 in Fig. 38). We suggest further that the faults that expose the Chinnur inlier (Fig. 15F) became active at this time. At some point within the interval between the Carnian Supra-Panchet/Dubrajpur Formations and the Early Cretaceous Rajmahal Trap/lamprophyre, the Damodar basins were deformed and uplifted by transpressional movement along the boundary faults and the northwest-trending cross faults. We regard this event as the same age (FA5) as the faulting in the Godavari area. Datta and Mitra (1984) and Mitra (1987, fig. 2) show right-lateral movement along the boundary faults of the Damodar basins, and Veevers et al. (1994a, fig. 3) illustrated 4.4 km of right-lateral offset in this area, as shown in Figure 44 in a later section. They also interpreted the structure of the Jharia basin as reflecting right-lateral movement. Mitra (1987, fig. 3) interpreted a right-lateral transpressional regime superimposed on the earlier extensional structure to generate uplift in the Son Valley area, shown by the parallelogram enclosing FA5 in India (Fig. 38).

According to Boote and Kirk (1989, p. 221), in earliest Jurassic time, the western and northwestern margins of Australia underwent "initiation of differential subsidence within extensional domains with wrench induced uplift/folding along incipient (major) transform offsets," which we show diagramatically on Figure 38 as right-lateral wrenching (FA5), as in India. The transform offsets are the Cape Range Fracture Zone, west of the Pilbara Block (PB), and the Wallaby-Perth Fracture Zone, southwest of Shark Bay (SB). This event is registered also in the Wombat Plateau of the northern Exmouth Plateau, near the edge

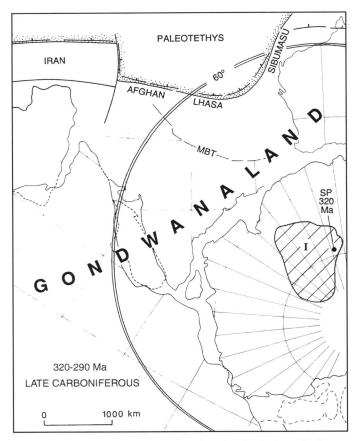

Figure 32. Regional paleogeography of Peninsular India and adjacent parts of Tethys and Gondwanaland from the mid-Carboniferous inception of Pangea through the late Carboniferous or Pennsylvanian (320 to 290 Ma). Asian blocks from Sengor (1987), including the Sibumasu block off Western Australia (Veevers, 1988a). The base map and paleolatitude are from Powell and Li (1994). Ocean floor is outlined by stipple. I denotes the ancestral Gamburtsev upland. MBT—Main Boundary Thrust; SP—South Pole.

of the map area, and involves post-Rhaetian major block-faulting with local uplift/subsidence/tilting . . . and major synrift volcanics (von Rad et al., 1989, fig. 13; 1992a, 1992b).

Right-lateral wrenching is pervasive in Pangea. It is registered initially in igneous rifts at 305 Ma and then at 290 Ma by differential subsidence in Europe and eastern Australia (Fig. 33B) (Veevers et al., 1994d, 1994f); subsequently by pre-Carnian folds in the northern Canning Basin (Fig. 38); and finally, starting in the earliest Jurassic and continuing to later Jurassic and Early Cretaceous breakup, by the structures in Australia and India mentioned above (Fig. 38). An exception to the sense of wrenching is the Permo-Carboniferous and Early Triassic *left*-lateral trantensional stress-field in East Africa (Wopfner, 1993).

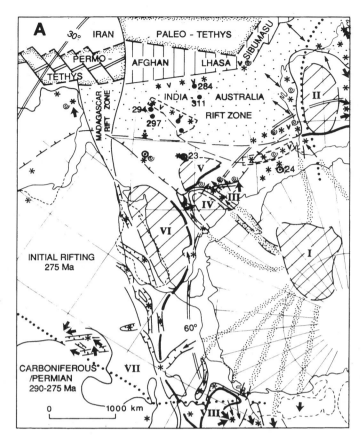
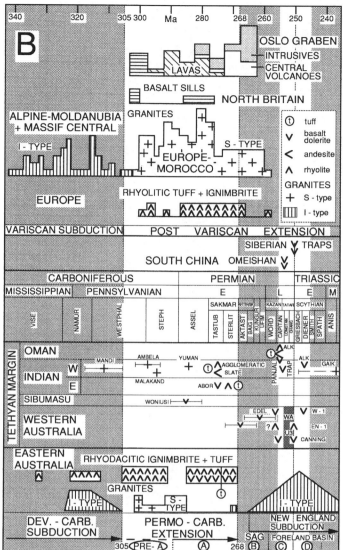

Figure 33. A, About the Carboniferous/Permian boundary (290 to 275 Ma), from Veevers et al. (1994c, fig. 5). The base map and paleolatitude are from Powell and Li (1994). Additional data on extra-Peninsular India, from Figure 1, include those on the East Himalaya from Srivastava et al. (1988) and on the rest of the Himalaya from Brookfield (1993, fig. 15), with the diamictite and overlying pyroclastics of the Agglomeratic Slate of Tastubian age (Kapoor and Takuoka, 1985, p. 33; Archbold, 1982) shown in their present position (boxed) north of the eastern part of the Main Boundary Thrust (broken line) and restored to their prethrusted position. Asselian glaciomarine Dingjiazhai Formation and overlying basalt of the Woniusi Formation of Yunnan (Xiaochi, 1994) placed near southwestern boundary of Sibumasu. Additional data on glaciogenic sediment (asterisk) and glacial flow in northwestern Australia are from O'Brien and Christie-Blick (1992); in Arabia from Braakman et al. (1982), including the probable Sakmarian (Tastubian) age of the postglacial marine sediment (Besems and Schuurman, 1987; Levell et al., 1988); and in central Africa from Frakes and Crowell (1970, fig. 8), with major faulting in East Africa and Madagascar at the end of glaciation (275 Ma) from Wopfner (1993, fig. 2). Additional data on structure about Paleotethys include the growth of Permotethys and the failed arm of the Madagascar rift zone (Stampfli et al., 1991) that followed the crustal grain of the Late Proterozoic Mozambiqean foldbelt. The marine shoreline (barred line) encloses occurrences of marine fossils (coils). The encircled dots indicate subsurface occurrences. Volcanic rocks are denoted by v's and and granitic rocks by filled circles, with ages in Ma. The initial rift-faulting in East Africa corresponds to the Tastubian (275 Ma) top of the Talchir Formation. Inferred ice flow from the ancestral Gamburtsev Mountains (I) is shown by lines of stipple, measured ice flow (from striations) by squat arrows, and fluvial flow by narrow arrows. Zambezian (fault-affected) terrain delimited by the heavy dotted line. Parallel barbed lines indicate grabens. Ocean floor is outlined by stipple. Uplands are—I ancestral Gamburtsev Mountains, II—ancestral Great Western Plateau, III—Chotanagpur, IV—Chhattisgarh, V—Faizabad, VI—Madagascar–southwest India, VII—Congo, VIII—Cargonian. B, Diagrammatic histograms of radiometric and biostratigraphic age determinations (not relative volumes) of various igneous rocks in the Oslo Graben, basalt sills in north Britain, and rhyolitic tuff/ignimbrite and granite from Europe and eastern Australia (from Veevers et al., 1994f), basalt from Siberia (Renne and Basu, 1991; Campbell et al., 1992) and South China (Zunyi et al., 1986) and from the Tethyan margin in Oman, the Indian subcontinent, Sibumasu and Western Australia (Veevers and Tewari, 1995). The timescale is modified from Palmer (1983) such that the Permian/Triassic boundary is taken to be 250 Ma (Claoué-Long et al., 1991). Pre-A, A–D refer to stratigraphic-tectonic stages of eastern Australia (Veevers et al., 1994e).

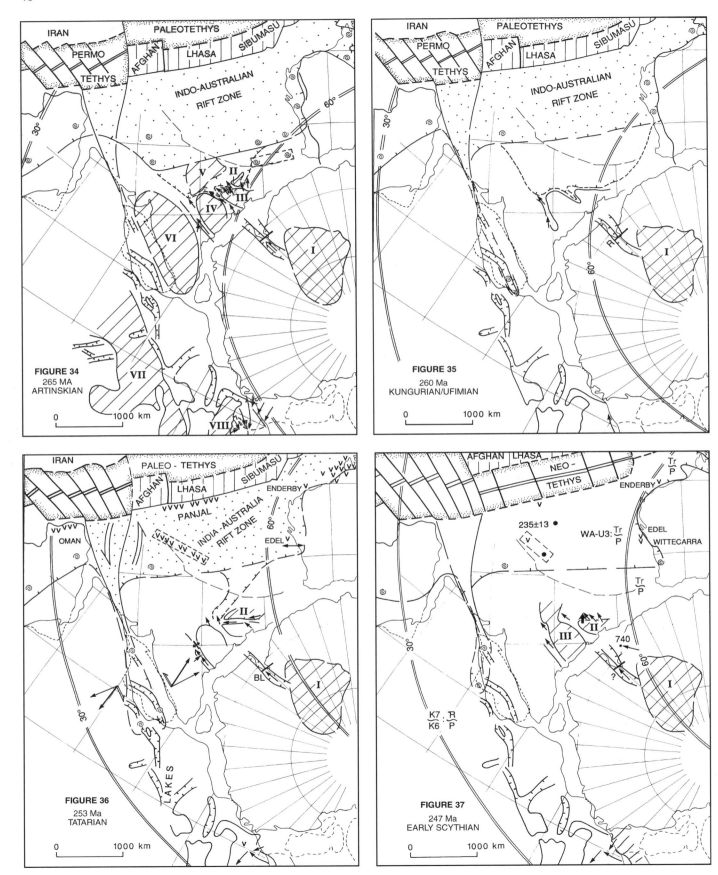

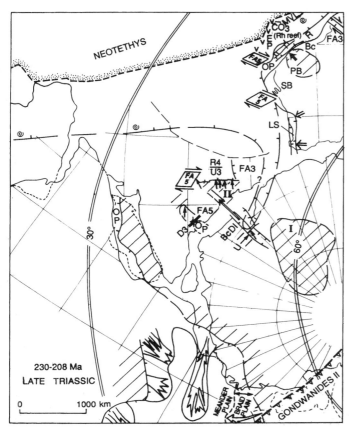
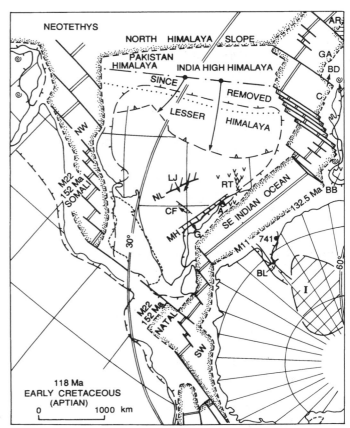

Figure 38. Late Triassic (230 to 208 Ma). The base map and paleolatitude are from Powell and Li (1994). I—ancestral Gamburtsev Mountains, II—Chotanagpur upland. EP—Exmouth Plateau, LS—Leseuer Sandstone, MVL—Mount Victoria Land (Veevers, 1988a), OP—Onslow Palynoflora, PB—Pilbara Block, R—Rankin delta, Rh—Rhaetian, SB—Shark Bay. Bc—*Bartenia communis*, DI—*Dicroidium* flora, CO_3—carbonate reef. Ocean floor is outlined by stipple.

Figure 39. Early Cretaceous–Aptian (M0 = 118 Ma). The oceanic reconstruction and paleolatitude are from de Wit et al. (1988) and Veevers et al. (1991). Palinspastic northern India and Pakistan are from Brookfield (1993, fig. 12). The shoreline in west India is from Robinson (1967), in Arabia from Murris (1980), and in southern Africa from Dingle et al. (1983, fig. 98). Abyssal Plains off Australia are AR—Argo, C—Cuvier, GA—Gascoyne. The V in southwestern Australia denotes the Early Cretaceous Bunbury Basalt (BB) (Veevers, 1984, p. 196) and the V in the Beaver Lake area the 110 ± 3-Ma Beaver Lake lamprophyre (BL) (McKelvey and Stephenson, 1990). The V's in India denote the Rajmahal Trap (RT). BD—Barrow Delta, NL—Narmada lineament. I—ancestral Gamburtsev Mountains. Ocean floor is outlined by stipple. Other symbols are explained in the text.

Figure 34. Artinskian (265 Ma). Symbols as in Figure 33A. The ocean floor is outlined by stipple. The base map and paleolatitude are from Powell and Li (1994). Uplands are I—ancestral Gamburtsev Mountains, II—Monghyr-Saharsa, III—Chotanagpur, IV—Chhattisgarh, V—Faizabad, VI—Madagascar–southwest India, VII—Congo, VIII—Cargonian, Witwatersrand Arch. Western Australia from BMR Palaeogeographic Group (1990, fig. Permian 4); Arabian carbonate and evaporites from Murris (1980).

Figure 35. Kungurian/Ufimian (260 Ma). The base map and paleolatitude are from Powell and Li (1994). Western Australia from BMR Palaeogeographic Group (1990, fig. Permian 5). R—Radok Conglomerate, the first deposit at Beaver Lake. Marine fossils in the Vohitolia Formation of Madagascar from Boast and Nairn (1982). I—ancestral Gamburtsev Mountains. Ocean floor is outlined by stipple.

Figure 36. Tatarian (253 Ma), shortly before the Permian/Triassic boundary (250 Ma). The base map and paleolatitude are from Powell and Li (1994). The Panjal Trap in its present position (boxed V's) is restored to its original position immediately south of the Afghan and Lhasa blocks. Other widespread volcanism behind Tethys included basaltic trap and other intraplate rift volcanics in the shelf succession of Oman and undersaturated volcanics (Edel), rhyolite (Enderby), and dolerite in Western Australia. The transtensional stress-field in East Africa and Madagascar (Wopfner, 1993) and east of Oman is explained in the text. The growing anticline in central India is shown by the anticlinal symbol. The double-headed arrow south of the Edel volcanics indicates extension. Uplands are I—ancestral Gamburtsev Mountains and II—Monghyr-Saharsa. BL—Beaver Lake. Ocean floor is outlined by stipple.

Figure 37. Early Scythian (247 Ma), shortly after the Permian/Triassic boundary (250 Ma). The base map and paleolatitude are from Powell and Li (1994). Uplands are I—ancestral Gamburtsev Mountains, II—Chotanagpur, III—Chhattisgarh. 740—Ocean Drilling Program drill-hole 740. Ocean floor is outlined by stipple. (Other symbols explained in text.)

Early Cretaceous, Aptian (M0 = 118 Ma)

Mount Victoria Land (MVL in Fig. 38) split off northwestern Australia in the Late Jurassic (M26, Oxfordian, ≈160 Ma, Veevers, 1988a; Veevers et al., 1991) by the earliest seafloor spreading of the Indian Ocean in the Argo Abyssal Plain (horizontal lines near AR [Argo] in Fig. 39). Greater India–Madagascar was isolated by seafloor spreading along a transform fault from M22 (152 Ma) in the northwest (NW, Somali Basin) and southwest (SW, Natal Basin) and from M11 (132.5 Ma) in the southwest and southeast (SE) (Fig. 39). Water from the new ocean basins lapped the newly formed continents in shallow peripheral seas, though off Antarctica direct evidence by marine sediment is unknown. An uplift along the continent-continent transform fault between the Cuvier (C) and Gascoyne (GA) areas shed sediment northward in the Barrow Delta (BD) (Veevers and Powell, 1979; Boote and Kirk, 1989).

Uplift along the Narmada lineament (NL) shed the Late Jurassic (LJ) alluvial-fan Bagra Formation and the Early Cretaceous Jabalpur Formation to the south across the former Chhattisgarh upland in a local reversal of the long-lived northwest paleoslope. The final evidence of the northwest paleoslope is provided by the piedmont breccia of the Chikiala Formation (CF) at the foot of the boundary fault of the northwest Pranhita-Godavari basin. To the southeast, the Mailaram High (MH) rose as the shoulder of the rift-valley arch that shed the Athgarh Sandstone (A) and Golapilli Sandstone (G) into the rift valley in a configuration mirrored by the other side preserved in Western Australia (Fig. 40). After seafloor spreading, this second reversal of paleoslope became permanent with thermal subsidence of the southeastern margin. The northwesterly paleoslope of Gondwana India that had lasted at least from the Late Carboniferous—if not much earlier: from the Late Proterozoic (Casshyap et al., 1993b, fig. 5)—reversed definitively only from the mid-Cretaceous. The present *southeast*-flowing Pranhita, Godavari, and Krishna Rivers, the Mahanadi River, and *east*-flowing Koel and Damodar Rivers give their names to the Permian and Mesozoic lobes of the Gondwana master basin that drained to the *northwest* from the dominant uplift of the ancestral Gamburtsev Mountains. A complication arose in the northeast with the eruption of the Rajmahal Trap (RT) and possibly the Bunbury Basalt (BB) in southwestern Australia from the head of a hot spot traced subsequently by the Ninetyeast Ridge to the present-day vents of Kerguelen.

Early Cretaceous, probably Albian (≈105 Ma) (Truswell, 1991) coal-bearing fluvial sediment in Ocean Drilling Program drill-hole 741 in Prydz Bay, East Antarctica (Turner and Padley, 1991) was deposited after breakup on the conjugate margin.

CONNECTIONS BETWEEN INDIA, ANTARCTICA, AUSTRALIA, AND AFRICA

Radial drainage system about the ancestral Gamburtsev Mountains

Within the Gondwanaland province of Pangea (Fig. 40), the individual lobes of the Gondwana master basin occupied the central-peripheral part of the broad Zambezian (fault-affected) terrain (ZT) in a system of consequent radial drainage that focused on the ancestral Gamburtsev Subglacial Mountains at the core of an East Antarctic paleoupland. To the side were the Collie (C) basin in Western Australia, described in Figures 43 and 44 in a later section, and the Waterberg (W) and Lower Zambesi (LZ) basins in Africa. Upslope from India was the Lambert Graben, as exposed in the Beaver Lake (BL) area, and downslope was the Tethyan margin. On the other side of the East Antarctic paleoupland were the foreland basin and Gondwanides foldbelt along the convergent Panthalassan margin of Gondwanaland (Veevers et al., 1994c). A close-up (Fig. 41A) shows that the sector of the drainage system from southern Africa to southeastern Australia had an area of 10,000,000 km^2 and the sector along the Panthalassan margin an area of 7,500,000 km^2, with a total area of the system (within the dotted line of Fig. 41A) of 17,500,000 km^2 (Table 8).

Modern analogs are found in central-eastern Asia (Fig. 41B). The drainage system about the Himalaya-Tibet upland has an area of 15,000,000 km^2 and that about the Mongolian Plateau an area of 12,000,000 km^2 (Table 8). The Himalaya-Tibet upland is buoyed up by 60-km-thick shortened crust, and Veevers (1994b) argues that the similarly thick crust beneath the Gamburtsev Subglacial Mountains possibly originated by shortening of zones of weakness beneath an intracratonic basin by long-distance transmission of stress from the mid-Carboniferous Variscan collision of Laurussia and Gondwanaland.

Stratigraphic columns of the latest Carboniferous-Permian-Triassic sediment deposited in the radial drainage system (Fig. 42) show a common sequence of the Gondwana facies of (1) Carboniferous-Permian glacigenic sediment at the base succeeded by (2) Permian coal measures except in Tasmania, dominated by marine sediment of the Panthalassan margin, (3) the barren sandstone and shale of the Early-Middle Triassic coal gap, and (4) the return in the Late Triassic, after the Gondwanides II deformation and coeval faulting in India (FA3, Table 5), of coal measures (C. M.) or, in India, carbonaceous shale. The columns come from comparable epicratonic or distal foreland basins. The Indian column accumulated fastest, with 2,200 m of Permian sediment in the Bokaro and Raniganj areas (Fig. 17), compared with 700 m in the Karoo (in the distal part of the foreland basin), 1,350+ m at Collie, and 800 m in Tasmania; India has 1,700 m of Triassic sediment in the northwest Pranhita-Godavari area, compared with 400+ m in the distal part of the Karoo and 600+ m in Tasmania. The faster rates of accumulation in India are probably due to the fact that the local relief between upland and valley in the Gondwana lobes in the fault-affected (Zambezian) terrain is greater than in the wider Karoo-type terrain. In the major rift system of the Perth basin (Geological Survey of Western Australia, 1990), 2,000 m of sediment accumulated during the Permian and 3,000 m during the Triassic. In the proximal (foredeep) part of the foreland basin that was depressed under the overthrusting orogen, approximately 5,000 m of Permian and 5,000 m of Triassic sediment accumulated in the Karoo basin of southern Africa (Veevers et al., 1994d) and in the Bowen basin of eastern Australia (Veevers et al., 1994e), a rate two to three times faster than that in India.

Radial drainage system disrupted by Pangean rifting

During the Middle Triassic, deformation by faulting and folding in India (Table 5, U3) and nondeposition in much of Australia and East Antarctica were followed by rifting and concomitant deposition in India (R4) and in the northern Perth basin, now part of a rifted arch (Fig. 43), of >2,500 m of Lesueur Sandstone, including rare thin coal seams, reflecting a return to coal-forming conditions in the Late Triassic (Veevers et al., 1994a). Fission-track dates on apatite of 232 ± 5 Ma and 208 ± 4 Ma from Archean rocks in the Yilgarn Block near the Darling Fault (FT in Fig. 43A) suggest that rifting was accompanied by rapid denudation and reversal of drainage. A second phase of rifting, in the Early Jurassic time of U5 and FA5 (Table 5), was accompanied by deposition of the 2,500-m-thick Cockleshell Gully Formation. Later rifting, in the earliest Late Jurassic and earliest Cretaceous, generated the 3,000-m-thick Yarragadee Formation. Then came breakup by seafloor spreading at 132.5 Ma and the deposition on the breakup unconformity of the Bunbury Basalt and the marine Warnbro Group.

Changes in the deep crust during rifting have maintained the step at the Darling Fault between the axis of the Perth basin and the Yilgarn Block. A comparable step in India is the Mailaram High that constricts the southeastern part of the Godavari basin. The Early Cretaceous Golapilli Sandstone in the Krishna basin and the Athgarh Sandstone to the northeast (both solid black in Fig. 43A) lie on the southeast paleoslope of the Mailaram High, which we interpret as the foot of the axial scarp. The present regional southeastern slope was probably not achieved until the K/T boundary (66 Ma), as indicated by eastward flow of the Deccan Trap (Rajahmundry Traps) to the present coast at Narsapur from the source in the west (Jaeger et al., 1989; Baksi et al., 1994).

Comparison of Damodar basins with the Collie basin

The chain of Late Paleozoic and Mesozoic events produced basins of similar size and structure in the Damodar River area of India, exemplified by those of Jharia and Hutar, and in the Collie area of Western Australia (Fig. 44). In the Hutar basin, the Precambrian/Supra-Panchet boundary (Fig. 44) was displaced dextrally from A to A' and the Precambrian/Karharbari boundary from B to B' by 4.4 km after the Late Triassic (Mitra, 1987). In the Jharia basin (Fig. 44), the main movement along the Southern Boundary Fault is normal. We interpret the en echelon fold axes A, B, C, and D and the set of orthogonal normal faults as indicating a critical amount of right-lateral movement along the Southern Boundary Fault; the double-shafted arrow is dissected into compression (C) orthogonal to the fold axes and extension (E) orthogonal to the faults. Unique to the Jharia basin is the inferred originally 3,000-m-thick overburden (Mishra, 1986, 1991; Mishra et al., 1990) and the subsequent structure that led to its loss by erosion. The Collie basin (Fig. 44) resembles the Jharia basin in size, structure, and fill, but evidence for transcurrent faulting is weak. The synclinal axis is asymptotic to the northeastern boundary fault, consistent with left-lateral movement, but other fold axes and cross-cutting faults are unknown. The similarities could reflect the reaction to Pangean stress of a mechanically uniform Precambrian basement, chiefly granitic gneiss on the flanks (Naqvi and Rogers, 1987; Geological Survey of Western Australia, 1990).

From Collie (C) in Western Australia through the Gondwana basins to Tuli (T) in east Africa (Fig. 43B), the alluvium of the radial drainage system was deposited on Precambrian basement—Rust's (1975) Zambezian tectono-stratigraphic terrain—that fractured along faults during or after deposition or both. Outside this area is the flexed basement of the Karoo terrain, with the ovate Karoo (VR and UE) and Paraná (RB) basins in the southeast and the Cooper (COO) basin in the northeast.

PANGEAN TECTONICS AND STRATIGRAPHY

We summarize in Figure 45 the tectonic events that were common to all parts of Pangea (Veevers et al., 1994c, 1994d), including Peninsular India, and the depositional environments common in the Gondwana facies of the Gondwanaland province or southern Pangea (Veevers, 1988b, 1993).

(1) 320- to 290-Ma lacuna and glacials I

The lacuna (vertical ruled lines) is interpreted as being caused by thermal uplift of the Pangean platform during the

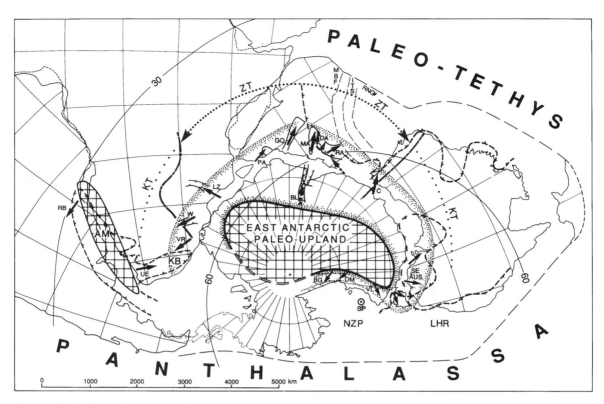

Figure 40. Radial drainage system in Gondwanaland about the East Antarctic paleoupland, from Tewari and Veevers (1993, fig. 2). This figure originally appeared in and is reprinted from: Findlay, R. H., Unrug, R., Banks, M. R., and Veevers, J. J., eds., 1993, Gondwana eight—Assembly, evolution, and dispersal—Proceedings of the eighth Gondwana Symposium, Hobart, Tasmania, Australia, 21–24 June 1991, 638 p. Reconstruction and paleolatitude of Early Permian southern Pangea (Gondwanaland) between Paleotethys and Panthalassa from Powell and Li (1994). The full arrows indicate the azimuth of Permian and Triassic fluvial transport and the broken arrows the azimuth in the Late Permian Raniganj Formation and Amery Group of Beaver Lake. Early Permian (Tastubian) marine ingressions are denoted by the heavy barbed line. Another paleoupland (cross-hatched) is the postglacial Atlantic Mountains (AM). BG—Beardmore Glacier, BL—Beaver Lake, C—Collie, DA—Damodar, DM—Darwin Mountains, GO—Godavari, ITS—Indus-Tsangpo Suture, KB—Karoo Basin, KT—basement of the Karoo (flexed) terrain (Rust, 1975), LHR—Lord Howe Rise, LZ—Lower Zambezi, MA—Mahanadi, MBF—Main Boundary Fault, NZP—New Zealand Plateau, PA—Palar, RA—Rajmahal, RB—Rio Bonito, RNGI—restored northern edge of Greater India, SE AUS—southeastern Australia, SP—South paleo-Pole, T—Tasmania, UE—Upper Ecca, VL—Victoria Land, VR—Vryheid, W—Waterberg, and ZT—Zambezian or fault-bounded terrain (Rust, 1975).

Figure 41. Large radial drainage systems, from Veevers (1994b, fig. 2). A, Early Permian southern Pangea with present-day 10° x 10° grid and South Pole (SP) and 280-Ma paleo–South Pole (SP 280 Ma) (Powell and Li, 1994), showing perimeter of radial drainage system (dotted line) and inferred first-order flow lines (stipple)—drawn back from flow indicators (arrows) to focus on the ancestral Gamburtsev Subglacial Mountains (GSM)—and perimeter of radial drainage about the central Australian uplands (heavy broken line). Bedrock contours (km) are indicated by light dotted and broken lines and crustal thickness (km) by the heavy broken line. The Bouguer gravity anomaly (*AB*, *CD*) is in mGal. Parallel lines indicate the structural trends of Proterozoic rocks along the coast of East Antarctica, Sri Lanka, India, and southwestern Australia and in the Prince Charles Mountains (PC). Late Permian fluvial channels in the Lambert Graben (L) trend northward. All other ice or fluvial flow azimuths pertain to the Early Permian, including Vestfjella (V) and the Amelang Plateau (AM), which has a semicircle of values about a southwest azimuth. Other abbreviations are AS—Alice Springs, BL—Beaver Lake, CO—Collie, EWM—Ellsworth-Whitmore Mountains, FI—Falkland Islands, GM—Grove Mountains, MG—Musgrave Block; PM—Petermann Ranges, W—Waterberg. B, Central Asia, at same scale as A, with radial drainage area (dotted line) about Himalaya-Tibet (>4 km elevation denoted by full line, >6 km by solid black), 60-km-thick crust (broken line), and drainage area (heavy broken line) about Mongolian Plateau (average elevation 2 km; dotted line denotes 3-km contour). Inset: At same scale, Rhine graben and Alps.

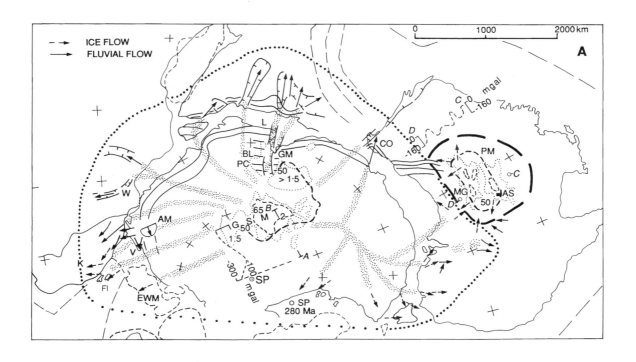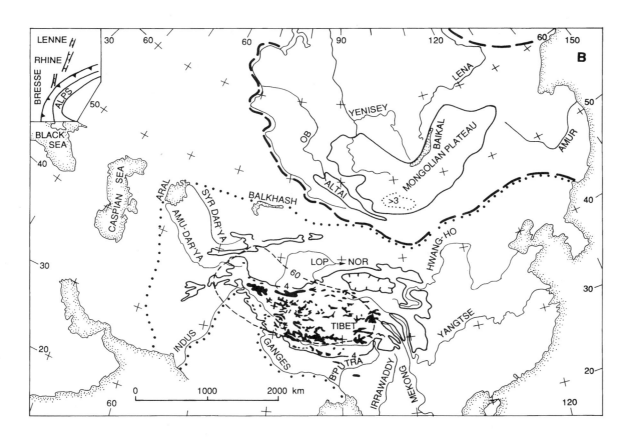

TABLE 8. LARGE RADIAL DRAINAGE SYSTEMS

Drainage System	Perimeter (km)	Length (km)	Area (km^2)
Modern Central-eastern Asia			
Himalaya-Tibet	17,000	500-2,500	15,000,000
Mongolian Plateau	13,000	250-2,500	12,000,000
Permian-Triassic Gondwanaland			
Panthalassan margin	5,000	1,500	7,500,000
Africa-India-Australia sector	10,000	2,500	10,000,000
		Total	17,500,000

accumulation of self-induced heat beneath the vast insulator brought together in the definitive collision of Laurussia with Gondwanaland (Veevers, 1989, 1990a). The Variscan collision belt (Veevers et al., 1994d) formed a high plateau at low latitudes. Andean-type foldbelts in eastern Australia and South America generated by subduction of Panthalassa also formed high plateaus, which became glaciated (Glacials I) because they were situated at high latitudes (Powell and Veevers, 1987). Deposition was confined to the oceanic margins of Pangea.

(2) 290-Ma Pangean extension I and glacials II

Extension of the platform provided a receptacle below base level for the accumulation of sediment at the base of the Pangean Supersequence, including the Gondwana facies of the Gondwanaland Province (Veevers, 1988b, 1990b). The age of 290 Ma is the numeric equivalent of eastern Australian palynologic stage 2, which dates the basal (glaciogenic) sediment of the Gondwanan basins (Glacials II). The Indian event is labeled R0a (Table 5). The last glaciogenic sediment in Antarctica and southern Africa was deposited before the Tastubian (275 Ma) marine transgression; in India glaciogenic deposition continued through but not beyond the Tastubian; and in eastern Australia it persisted to the end of the Permian.

(3) Coal I

Deposition of coal followed the regression of the sea in the Sterlitimakian (272 Ma) and continued with or without break to the end of the Permian in all the areas shown in Figure 45. Notable breaks were during the deposition of the Barren Measures in India and during the barren interval in southern Africa after the Ecca coal measures. The close of the Permian saw the end of coal deposition I except perhaps for a short overrun at Beaver Lake in East Antarctica. The antipathetic redbed facies is seen first in India at 260 Ma, a little later in southern Africa, and in the earliest Triassic elsewhere except along the Tethyan margin at the Salt Range, where redbeds started in Middle Triassic time. Farther north on the Tethyan margin, the Panjal Trap was erupted along a rift-valley complex that became the site of seafloor spreading that separated the Afghan, Lhasa, and Sibumasu blocks from the mainland. In Western Australia, alkaline volcanics beneath and above an angular unconformity (WA-U3) are interpreted as reflecting the Tethyan breakup.

(4) Coal gap

During the Early and Middle Triassic, the tectonic style of India was unchanged, but no coal or barely any other organic matter is known in India or elsewhere, except perhaps in New Zealand. Red pigment replaced black at the gross change in environment and biota at the Permian-Triassic boundary (Veevers et al., 1994b; Conaghan et al., 1994). In company with southern Africa and East Antarctica, all but the earlier part of the Triassic is a sediment gap or lacuna in Peninsular India during uplift and erosion.

(5) Gondwanides II (~233 Ma)

The Gondwanides II event involved the terminal deformation of the Samfrau Geosyncline and yoked foredeep (Du Toit, 1937), dated in South Africa by 230 ± 3 Ma metamorphic minerals in the Cape Fold Belt, with "final deformation of all pre-Beaufort rocks by kink bands and lower Beaufort rocks by listric thrusts" (Hälbich et al., 1983, p. 158). Gondwanides II is represented by a lacuna between the youngest (240 Ma) preserved folded Beaufort Group and the overlying flat Carnian (230 Ma) coal-bearing Molteno Formation; in eastern Australia, it is represented by the lacuna between the 234-Ma Moolayember Formation, the youngest deformed, and the 230-Ma Ipswich Coal Measures, deposited during extension of the foldbelt. Outside the foldbelt, transpressive folding in the Canning Basin of northwestern Australia is also dated as 230 Ma (Veevers, 1990b). Less narrowly dated is the Middle-Late Triassic final coalescence of Pangea that involved the accretion by col-

Figure 42 (on facing page). Composite columns of the latest Carboniferous-Permian-Triassic sediment deposited in the radial drainage system about the ancestral Gamburtsev Mountains, from the Karoo basin (KB in Fig. 40A), the Gondwana master basin, the Collie basin (CO in Fig. 41A), and the Tasmania basin. The meter scale is uniform except in the >3,000-m Gondwana column, at half scale. The Permian-Triassic (P/Tr) boundary is denoted by the dotted line. The Karoo column is a composite from Veevers et al. (1994d, fig. 8, I, III, VIII), the Gondwana column is a composite of the Jharia basin (Fig. 44 in a later section) and the Panchet Formation (Fig. 11, XXVIII), the Collie column is from Figure 44 in a later section, and the Tasmania column is from Veevers et al. (1994e, fig. 25, XXVI). C. M.—coal measures, PG—Pietermaritzburg.

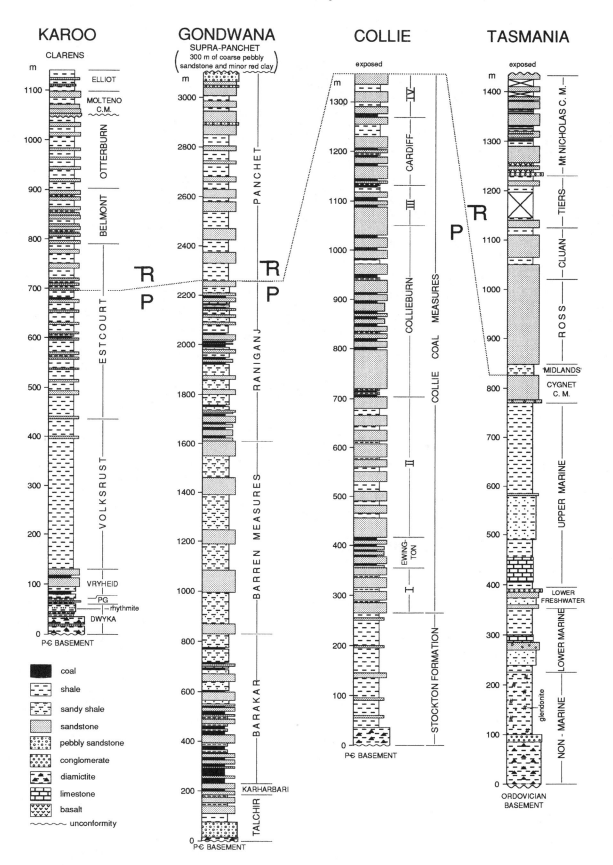

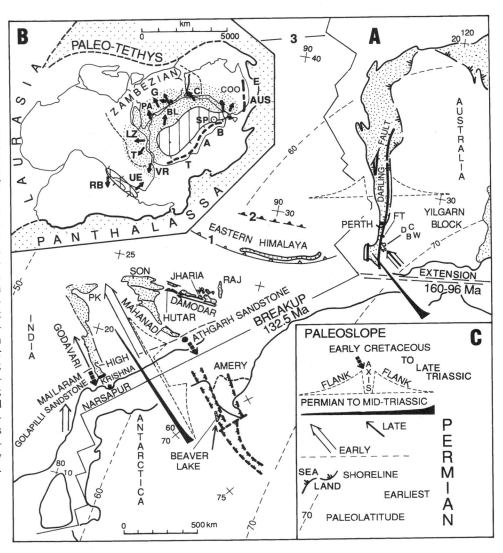

Figure 43. Disruption by rifting, from Veevers et al. (1994a). A, Permian-Jurassic reconstruction by eliminating continental extension and seafloor spreading between India, Antarctica, and Australia and crustal shortening north of Main Boundary Fault (1) and Indus-Tsangpo Suture (2) to restored northern edge of Greater India (3) (Powell et al., 1988). Exposed basins stippled (PK—Pench-Kanhan; RAJ—Rajmahal) except Australian cluster east of Darling Fault: Donnybrook (D), Collie (C), Boyup (B), Wilga (W); FT—fission-track ages (Ferguson, 1981). Heavy lines denote faults. B, Pangean reconstruction, with Permian fluvial transport azimuth or regional paleoslopes (heavy arrows) on flanks of East Antarctic upland. BL—Beaver Lake; C—Collie; COO—Cooper; E AUS—eastern Australia; G—Gondwana; LZ—Lower Zambezi; PA—Palar; RB—Rio Bonito; SP—South Pole; T—Tuli; TAB—Transantarctic Basin; UE—Upper Ecca; VR—Vryheid. C, key to paleoslopes and other features in A, including Permo-Carboniferous paleolatitudes from Veevers and Powell (1987) and Permian to mid-Triassic northwest slope disrupted by Late Triassic to Early Cretaceous rifting.

lision of South China and Cimmeria to the Paleotethyan margin (Veevers, 1990a). We suggest that Indian uplift (Table 5, U3), faulting (FA3) and other deformational events (D3), bracketed within the lacuna from 247 to 230 Ma, occurred at this time.

(6) 230-Ma (Carnian) onset of Pangean extension II and Coal II

The 230-Ma (Carnian) onset of Pangean extension II and Coal II are represented by the start of Pangean rifting, including the rift basins about the North Atlantic and Eastern Indian Oceans (Veevers, 1989, 1990a), the intramontane Ipswich Coal Measures in eastern Australia, the main coal measures in Tasmania, and the Leigh Creek coal measures in central Australia (Veevers, 1990b); the Molteno Formation during relaxation of the flanks of the Cape Fold Belt of southern Africa; coal in the Transantarctic Mountains; and carbonaceous sediment in New Guinea, East Antarctica, and Western Australia and in the Dubrajpur Formation (DUB in Fig. 45) of the Rajmahal basin of Peninsular India. The couplet of Gondwanides II–Extension II is reflected by U3-R4 in India. Redbeds in the Supra-Panchet Formation in the Koel-Damodar area coexist with dark beds at this level (Fig. 5), as in the upper Flagstone Bench Formation of East Antarctica; redbeds in southern Africa continue with the Elliot Formation.

(7) Earliest Jurassic (208 Ma) block-faulting by right-lateral transcurrence

The faulting events in India about the Triassic-Jurassic boundary (Table 5, FA5) are dated by the earliest Jurassic basal pebbly sandstone of the Kota Formation, which Mitra (1987,

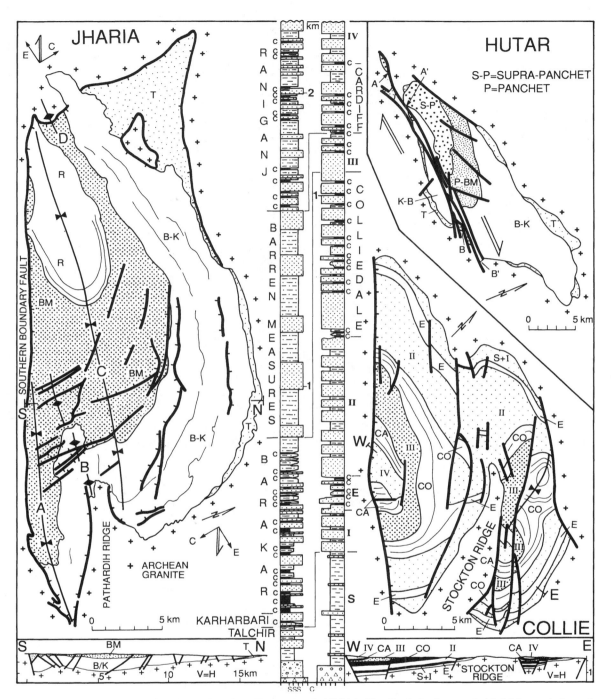

Figure 44. Jharia, Hutar, and Collie basins, from Veevers et al. (1994a). The Jharia basin map (with fold axes A through D and selected faults), cross section, and 2,240-m-thick stratigraphic column are from Ghosh and Mukhopadhyay (1985), supported by Mukhopadhyay (1984) and Mukhopadhyay et al. (1994). Letter symbols on the map are T—Talchir, B-K or K-B—Barakar-Karharbari, BM—Barren Measures, R—Raniganj. The profile of the stratigraphic column is scaled for CD—conglomerate/diamictite, SS—sandstone, SI—siltstone, SH—shale; coal environments (c—coal) are from Mitra (1991). The Hutar basin map, from Raja-Rao (1987), shows a 4.4-km right-lateral offset of the Supra-Panchet Formation from point A to A′ (Mitra, 1987) and of the base of Karharbari Formation from point B to B′. The letter symbols are the same as in the Jharia map. The Collie basin (C in Fig. 43A) map and 1,350-m-thick stratigraphic column, at nearly twice the scale of the Jharia column, are from the Geological Survey of Western Australia (1990). The Stockton Formation (S) is overlain by the Collie Coal Measures, in which I to IV denote noncoal intervals except two thin seams in III and E denotes the Ewington coals, CO the Collieburn coals, and CA the Cardiff coals.

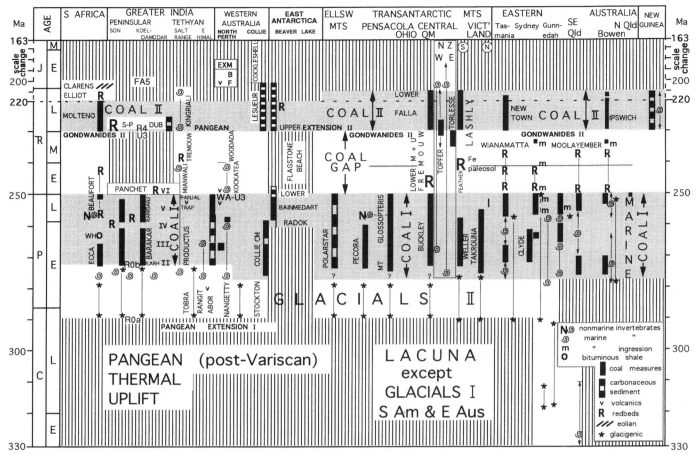

Figure 45. Timetable of Carboniferous to Jurassic tectonics and environments of Greater India and neighboring Africa, Australia, and Antarctica. Information about Greater India, Western Australia, and Antarctica has been added to fig. 3 of Veevers et al. (1994c). DUB—Dubrajpur Formation, EXM—Exmouth Plateau, FA5—earliest Jurassic faulting, KARH—Karharbari, R4—Carnian relaxation, S-P—Supra-Panchet Formation; U3—pre-Carnian uplift, WA-U3—latest Permian/earliest Triassic block faulting, WH—Whitehill Formation.

p. 36) interprets as a debris-flow deposit at the foot of a fault block. At the same time, the western and northwestern margins of Australia underwent differential subsidence with wrench-induced uplift and folding along incipient (major) transform offsets, including synrift volcanics in the northern Exmouth Plateau (EXM, v BF in Fig. 45).

(8) Early Cretaceous breakup of India from Antarctica and Australia

Continued rifting between Greater India and Australia changed at 132.5 Ma (mid-Valanginian) into the seafloor spreading that finally isolated India.

ACKNOWLEDGMENTS

Our collaboration was made possible by grants under the Indian-Australian Science Agreement. We are grateful to the governments of India and Australia for facilitating Tewari's visit to Macquarie University in April to June 1989 and Veevers's return visit to Aligarh in January and February 1990. Tewari's visit to Macquarie in February 1994 was supported by a grant to Veevers from the Australian Research Council, which has supported Veevers's work in Gondwanaland for the past 20 years.

We thank the Head of the School of Earth Sciences, Macquarie University, and the Head of the Department of Geology, Aligarh Muslim University, for the provision of facilities. In India, we visited H. K. Mishra at the Central Mine Planning and Design Institute Limited, Ranchi, and in the field, and Veevers visited the Oil and Natural Gas Commission, Dehra Dun, at the invitation of S. K. Biswas, and the Wadia Institute of Himalayan Geology, Dehra Dun, at the invitation of V. C. Thakur. We had valuable discussions with N. D. Mitra in Aligarh during January 1990.

We acknowledge the help of E. J. Cowan in making a preliminary set of references; P. J. Conaghan, R. Morgan, and Xiaochi Jin for supplying references; Judy Davis for the final line-drawings; and Adam Bryant and I. G. Percival for the computer drawings. R. Helby provided valuable advice on palynology.

We are grateful to the GSA reviewers, S. M. Casshyap, P. E. Potter, and J. J. W. Rogers, for the prompt effort they put into their constructive criticism of the manuscript and to N. D. Mitra, whose brief comments survived the postal system that mislaid his expanded comments.

Figure 3 was adapted from Naqvi and Rogers (1987) with permission of the senior author and the publisher (Oxford University Press). Figure 40 was adapted from Tewari and Veevers (1993) with permission of the publisher (A. A. Balkema).

APPENDIX 1. NOTES ON TIME-CORRELATION OF THE GONDWANA FORMATIONS

The biostratigraphic timescale in Figures 4, 5 and 6 is calibrated in millions of years by the DNAG timescale (Palmer, 1983) and modified such that the Permian-Triassic boundary is dated as 250 Ma, as in Veevers et al. (1994b).

Correlation between India and elsewhere

The chief means of time-correlation of the Gondwana formations is through Australian palynomorphs and invertebrates, supplemented through vertebrates from other fragments of Gondwanaland. The numbered points that follow are encircled (in miscellaneous order) on Figures 4 and 5.

1. The Tobra Formation of the Salt Range (Fig. 5) is correlated with East Australian Palynologic Stage 2 (Truswell, 1980, p. 99).

2. The Narmia Member is within the Scythian-Anisian *Tigrisporites playfordii* zone (Foster, 1982, p. 176).

3. The Mittiwali and Kathwai Members are within the *Protohapoloxypinus samoilovichii* and *Lunatisporites pellucidus* zones (Foster, 1982, p. 176).

4. The upper part of the Chhidru Formation, within the *Playfordiaspora crenulata* zone (Foster, 1982, p. 176), is dated uppermost Permian (Changsingian) by the marked drop in $\delta^{13}C_{CO3}$ at the Permian-Triassic boundary (Baud et al., 1989). In the Bowen basin, Queensland, the *P. crenulata* zone occupies the uppermost 8 m of the coal measures and is overlain by the *Protohapoloxypinus microcorpus* zone at the base of the barren Rewan Group, marked by a drop in $\delta^{13}C_{org}$ that indicates the Permian-Triassic boundary (Morante et al., 1994).

5. Spores in the Tredian Formation indicate a Middle Triassic age (Balme, 1970, p. 423).

6. The Talchir Formation in the South Rewa (Umaria) and Jainty coalfields is equivalent to East Australian Palynologic Stage 2 (Kemp, 1975, p. 406).

7. The microflora of the Talchir Formation is correlated with the *Granulatisporites confluens* zone (Foster and Waterhouse, 1988, p. 146).

8. The Bap Formation of Rajasthan contains the *G. confluens* zone (Foster and Waterhouse, 1988, p. 147).

9. (a) The Raniganj Formation (Fig. 4) is equivalent to East Australian Palynologic Upper Stage 5 (Foster, 1979, p. 126). (b) The Maitur Formation (Maheshwari and Banerji, 1975), corresponding to the lower part of the Panchet Formation, is equivalent to the *P. microcorpus* zone (Foster, 1979, p. 126), which Morante et al. (1994) determine as earliest Triassic.

10. The lower part of the Barakar Formation is tentatively equivalent to East Australian Palynologic Stage 4 (Foster, 1979, p. 126).

11. The Karharbari Formation is tentatively correlated with the Sterlitamakian-Aktaskian Sardi-Warchha Formations of the Salt Range (Fig. 5) and the Poole Sandstone of the Canning basin, Western Australia, equivalent to East Australian Palynologic Stage 3b–Lower 4 (Balme, 1980, p. 52).

12. (a) The Dubrajpur Formation (Fig. 4) contains palynomorphs that indicate Carnian (Tiwari and Tripathi, 1987; Helby et al., 1987; R. Helby, personal communication, 1989). (b) Palynologic assemblage zones D, C, B, and A in Kommugudem-A well in the Krishna-Godavari basin (Fig. 5) contain the Onslow Palynoflora, equivalent to Dolby and Balme's (1976) Carnarvon basin zones *Tigrisporites playfordii* to *Minutosaccus crenulatus* (Prasad and Jain, 1994), equivalent to Spathian through Norian.

13. Concerning vertebrate zones in the northwest Pranhita-Godavari basin: (a) Vertebrates in the lower part of the Kamthi Formation indicate the *Tapinocephalus* zone of South Africa (Anderson, 1981, p. 6, 7), equivalent to East Australian Palynologic Upper Stage 5 b/c or PP 5.2. (b) Vertebrates near the base of the middle part of the Kamthi Formation indicate the all-but-latest Permian *Cistecephalus* zone of South Africa (Tiwari and Vijaya, 1992, p. 38). (c) (i) The Kamthi Formation of the Wardha coalfield contains vertebrates that indicate the latest Permian *Daptocephalus* zone (Satsangi, 1988, p. 248). (c) (ii) The Hirapur (upper Panchet) Formation or Deoli Stage of the North Karanpura and Raniganj basins and the Mangli Beds of the Wardha Valley belong to the early Scythian *Lystrosaurus* zone (Satsangi, 1988, p. 248). (d) The Yerrapalli Formation has a fauna roughly equivalent to the late Scythian *Cynognathus* zone (Jain and Roychowdhury, 1988, p. 222). (e) The Maleri Formation contains a Carnian fauna (Jain and Roychowdhury, 1988, p. 222). (f) The Dharmaram Formation contains a late Norian and Rhaetian fauna (Jain and Roychowdhury, 1988, p. 222).

14. The palynoflora of the lower (coal-bearing) part of the Kamthi Formation resembles the Raniganj palynoflora (Srivastava and Jha, 1988, p. 123).

15. We place the boundary between the Barren Measures and the Barakar Formation halfway through East Australian Palynologic Stage 4, calibrated as 260 Ma (Figs. 4 and 5).

16. Palynozones I to VI (lower half of Figs. 4 and 5) are from Tiwari and Tripathi (1988).

17. Leiosphaerids associated with Tastubian invertebrates (Venkatachala and Tiwari, 1988) are presumed to be within Palynozone IB.

18. Marine invertebrates: (a) Salt Range (Fig. 5): according to Dickins and Shah (1981), the Dandot Formation is Tastubian and the Amb Formation Baigendian-Kungurian; the Wargal Formation is not older than Kazanian. The Chhidru Member is Changsingian (Haq et al., 1988, p. 91; Baud et al., 1989), and the Mianwali Formation is Scythian (Haq et al., 1988, p. 91). (b) According to Archbold (1982), brachiopods in the Bap Formation and (c) in the Talchir Formation at Umaria indicate late Tastubian; (d) and at Manendragarh, they indicate

early Tastubian. (e) In the Talchir Formation at Rajhara and Daltonganj, invertebrates indicate late Tastubian (Dickins and Shah, 1979, p. 389). (f) The Talchir Formation at Hutar contains foraminiferans (Chaudhuri, 1988). (g) In the Raniganj and Jharia coalfields, the Raniganj Formation contains bivalves that are nonmarine except for two genera that have "marine" affinities (Chandra and Betakhtina, 1990; Chandra, 1994).

19. Trace fossils in the upper part of the Talchir Formation of the Raniganj and Deogarh basins have a marine aspect (Guha et al., 1994).

20. Foraminiferans were reported by Pal et al. (1994) in (a) the Talchir Formation of the West Bokaro coalfield and (b) the Kulti Formation (= Barren Measures) of the Raniganj Coalfield (Fig. 4).

* * *

Ages of formations in the Tethyan Himalaya (located on Fig. 1) come from Kapoor (1992, p. 24), supported by Acharyya et al. (1979).

Correlation within India, with notes on thickness and the environment of deposition

Tiwari and Tripathi (1988) established a succession of palynofloras or compositions on the basis of relative abundance of morphotype groups (I A and B, etc., in Figs. 4 and 5; Table 2).

Reflecting changes in paleoclimate in Peninsular India, the palynofloras indicate that the boundaries between the formations are synchronous and therefore confirm Pascoe's (1968, p. 923–1016) view that the formations are also stages.

Venkatachala and Tiwari (1988) report leiosphaerids (L) and phosphorite (PO_4) as possible indicators of marine incursions, principally in the Talchir Formation and Barakar Formation. Leiosphaerids are commonly associated with marine invertebrates, but as they range from marine to brackish environments, they are not by themselves diagnostic (R. Helby, personal communication, 1990).

Time-correlation and environmental features are now documented for the following areas as illustrated in the time-space diagrams (Figs. 4, 5, 16) and the thickness diagrams (Figs. 17 and 18). We start in the northwest.

Salt Range. Located northwest of the Peninsula (Fig. 1), the data from the Salt Range (Fig. 5), in particular the start of reddish coloration, are from the Pakistani-Japanese Research Group (1985); time-correlation is documented above.

Bap-Jaisalmer area. Located also to the northwest (Fig. 1), the outcropping Permian Badhaura and Bap Formations (Fig. 5) and the Late Proterozoic–Early Cambrian Nagaur Formation are from Ranga-Rao et al. (1979, p. 490); the subsurface data are from Datta (1983). Leiosphaerids (L) in the Bap Formation are from Venkatachala and Tiwari (1988), and the *G. confluens* palynologic zone is from Foster and Waterhouse (1988). Invertebrates in the Bap Formation indicate a Tastubian age, equivalent to East Australian Palynologic Stage 3a, and in the Badhaura Formation an Aktaskian-Sterlitimakian age, equivalent to Stage 3b (Archbold, 1982).

Umaria. The formations (Fig. 5) and their thicknesses are from Raja-Rao (1983, p. 120). The microflora of the Talchir Formation in the South Rewa Coalfield is equivalent to East Australian Palynologic Stage 2 (Kemp, 1975, p. 406), and the age of the invertebrates is late Tastubian (Archbold, 1982, p. 290). Biozonal details of the Karharbari Formation in this and other Peninsular areas are from Banerjee (1988). According to Mitra (1993b), the Pali Formation is at least 500 m thick and comprises a lower 300 m of red-brown clay and sandstone, with carbonaceous shale; a middle 200 to 250 m of green to reddish shale and white sandstone with interbanded coals up to 3 m thick, with a Raniganj flora; and an upper interbedded chocolate to green shale and sandstone, with a flora of Late Permian affinity and a palynoflora with a Permian/Triassic transitional aspect or, according to Tiwari and Ram (1986), a Late Permian aspect; the upper part probably extends through the Early Triassic. The Tiki Beds contain reptiles of the same (Carnian) age as those in the Maleri Formation (Anderson, 1981, p. 6;

Kutty et al., 1988, p. 225; Sengupta, 1992) and a Late Triassic microflora (Sundaram et al., 1979). The Parsora Formation has a gradational and conformable contact with the underlying Pali Formation (Sastry et al., 1977, p. 70–71) and accordingly is placed in the Middle Triassic (Mitra, 1993b). Alternatively, from its Late Triassic (?Rhaetic) flora, the Parsora Formation can be placed higher. We show both possibilities in Figure 5. Its relations are discussed further by Dutta and Ghosh (1993). The Cretaceous/Tertiary (K/T) Deccan Trap, the Lameta Beds, and the Early Cretaceous Jabalpur Formation overlap all the older formations (Robinson, 1967, p. 244).

Singrauli. The main data about the Singrauli area (Fig. 4) are from Raja-Rao (1983, p. 132); thicknesses of formations, except the Panchet and Supra-Panchet, are unknown. Leiosphaerids in the Barakar and Raniganj Formations are from Venkatachala and Tiwari (1988). In the Raniganj Formation, the upper (top) Jhingurdah Seam is 105.00 to 159.70 m thick (Fig. 15E) but contains up to 47% of dirt bands (Casshyap and Tewari, 1988, p. 62; Raja-Rao, 1983, p. 143). The Nidpur Beds are bounded by faults, and the estimated age depends entirely on fossils, mainly palynomorphs, which are of Middle Triassic age (Srivastava, 1988, p. 159).

Ramkola. The Supra-Panchet Formation (Fig. 4) is 250 m thick, the Panchet Formation 350 m, the Raniganj Formation 300 m, the Barren Measures Formation 150 m, the Barakar and Karharbari Formations together 600 m, for a grand total of 1,650 m (Casshyap and Tewari, 1984, p. 125).

Daltonganj. The main data (Fig. 4) come from Raja-Rao (1987, p. 270). The marine invertebrates in the Talchir Formation at Rajhara, near Daltonganj, are dated as Tastubian (Dickins and Shah, 1979, p. 389; Archbold, 1982, p. 270).

Hutar. Details (Fig. 4) are again from Raja-Rao (1987, p. 258). The foraminiferans at the Talchir/Karharbari boundary are from Chaudhuri (1988).

North Karanpura. The details (Fig. 4) are from Casshyap and Kumar (1987, p. 183). According to Pascoe (1968, p. 1030), "the Raniganj Beds overlap the Ironstone Shales [Barren Measures] and rest on the Barakars." In a detailed map (Raja-Rao, 1987, plate XII), this boundary is shown as a fault. The clear angular unconformity between the Supra-Panchet and Panchet Formations is shown in Fig. 15D.

Bokaro. According to Pascoe (1968, p. 1027), "The Raniganj stage . . . is found . . . resting on the Ironstone Shales [Barren Measures] with a well marked unconformity and abruptly overlapping them in places. . . . Their unconformity [of the Panchets] to the Damudas is striking . . . they completely overlap both the Raniganj and Ironstone Shales . . . and rest upon Barakar beds." Raja-Rao (1987, plates VII and VIII) again interprets this "unconformity" as a fault. But, higher in the succession, "This sequence . . . tentatively correlated with the Mahadeva [Supra-Panchet] Formation . . . rests over the Panchet Formation with an apparent angular unconformity and this unconformable junction can be observed on the eastern face of Lugu Hill" (Raja-Rao, 1987, p. 12). The leiosphaerids in the Talchir Formation (Fig. 4) are from Venkatachala and Tiwari (1988).

Jharia. The Raniganj Formation (Fig. 4) is the youngest preserved formation. We outline the history of uplift and the stripping of the younger succession in the text accompanying Figure 44. Leiosphaerids in the Talchir Formation are from Venkatachala and Tiwari (1988).

Deogarh. The chief source of information is Tewari (1980). The Deogarh region (Fig. 4) contains two coalfields, 40 km apart, in which the formations have these thicknesses:

	GIRIDIH	SAHARJURI-JAINTI (S-J)
	(m)	(m)
Barakar	180	300
Karharbari	210	350
Talchir	91	250

In the Jainti area, the microflora of the Talchir Formation is equivalent to East Australian Palynologic Stage 2 (Kemp, 1975, p. 406).

Raniganj. The stratigraphic details (Fig. 4) are from Casshyap and Kumar (1987, p. 183). Phosphorite (PO_4) in the Raniganj Formation is from Venkatachala and Tiwari (1988). The Supra-Panchet Formation rests unconformably on the folded Panchet Formation.

Rajmahal. Details of the Permian succession (Fig. 4), which reaches no higher than the Barakar Formation, are from Raja-Rao (1987, p. 303–307). The Talchir Formation ranges up to 100 m thick and, as in the Koel-Damodar areas, overlies Precambrian rocks; leiosphaerids in the Talchir Formation of the Chuperbhita Coalfield are from Venkatachala and Tiwari (1988). The Barakar and Karharbari Formations together range from 250 to 550 m thick.

The disconformably overlying Dubrajpur Formation contains carbonaceous shale, very rare at this level, that yielded palynomorphs of Ladinian-Carnian age (Tiwari and Tripathi, 1987) by comparison with a form found also in the Sahul Shoals-1 Well in northwest Australia (R. Helby, personal communication, 1989). We show this stratigraphic range in Figure 4 but place the Dubrajpur Formation wholly within the Carnian at the top of the range, at the same level as the Supra-Panchet Formation. The Dubrajpur Formation is 60 to 250 m thick and overlaps the Permian succession to rest on Precambrian basement (Fig. 15A). Older Triassic rocks are found in the subsurface to the east and are dated by palynomorphs as Middle Triassic (Raja-Rao, 1987). The eastern side of the Rajmahal area is covered by the Rajmahal Trap, 600 m of flow basalt, 120 to 110 Ma old (Baksi et al., 1987) and inter-trappean sediments with a Jurassic/Cretaceous flora. The Rajmahal Trap overlaps older formations to rest on the Precambrian (Fig. 15A).

Singrimari. The Singrimari area, located on Figure 1, contains a succession (Fig. 4) of 75 m of the Talchir Formation, 110 m of the Karharbari Formation, and probably, in the subsurface to the west, the Barakar Formation (Srivastava et al., 1988, p. 331, 334).

Manendragarh-Sohagpur. We classify the column (Fig. 5) in terms of formations in the Koel-Damodar area, though as Raja-Rao (1983, p. 101, 102) points out, the formations are not precisely the same. The invertebrates in the Talchir Formation are early Tastubian (Archbold, 1982, p. 270).

Korba/Mand-Raigarh. According to Raja-Rao (1983, p. 15, 22, 24), the column (Fig. 5) is similar to that of the Sohagpur area up to the Barakar Formation and phosphatic Barren Measures (Datta, 1986a, p. 2). At the top of the column, the Raniganj Formation, with thin coal, disconformably overlaps the Barakar Formation.

Ib River. According to Raja-Rao (1982, p. 53), the concordant succession (Fig. 5) comprises 130 m of the Talchir Formation, 125 m of the Karharbari Formation, 600 m of the Barakar Formation, and Barren Measures, with leiosphaerids and phosphate nodules (Datta, 1986a, p. 2; Venkatachala and Tiwari, 1988, p. 27), all overstepped by the reddish conglomerate, sandstone, and shale of the Kamthi Formation (without coal) (Raja-Rao, 1982, p. 54), which together with the Barren Measures is 600 m thick (Mukhopadhyay et al., 1984, p. 558).

Talcher. The data come from Raja-Rao (1982, p. 42, 44). Mitra et al. (1979a, p. 38) give the names and thicknesses of the formations, including supporting information on the lithologic identification of the Supra-Panchet Formation (Fig. 5). According to Datta (1988), "Mapping in southern part of the western half of Talcher Coalfield reveals that the metamorphics are directly overlain by the Kamthis [Raniganj] across the boundary fault." On the east, the Talchir Formation (Venkatachala and Tiwari, 1988) is unconformably overlain by the Cretaceous Athgarh Sandstone (Kumar and Bhandari, 1973).

Satpura. The data (Fig. 5) come from Raja-Rao (1983, p. 157) and Peters (1991). The Bijori Formation, with Late Permian vertebrates and a microflora (Raja-Rao, 1983, p. 159) of Raniganj age, is disconformably overlain by the type Mahadeva Group, comprising the Pachmarhi Sandstone and the reddish Denwa Formation, with vertebrates correlated with the Yerrapalli Formation of the northwest Pranhita-Godavari basin (Satsangi, 1988, p. 249). Also originally included in the Mahadeva Group was the Bagra Formation, but Casshyap et al. (1993a) argue that it is better placed at the base of the following Jurassic-Cretaceous succession (Fig. 6). Following a Permian-Triassic northward paleoslope, the Bagra Formation was deposited in south-directed alluvial fans at the foot of the uplifted Early Proterozoic Bijarwar Supergroup and oversteps all older formations, as do the succeeding Jabalpur Formation, Lameta Formation, and Deccan Trap.

Some authors (e.g., Sastry et al., 1977) have extended the term "Mahadeva" to the Koel-Damodar and Mahanadi areas, but we prefer the term "Supra-Panchet" for the formation that overlies the Panchet Formation.

Kamptee. The data (Fig. 5) are from Raja-Rao (1982, p. 75–77). We interpret the reddish barren Kamthi Formation here and in the Wardha area as equivalent to the middle and upper parts of the Kamthi Formation of the northwest Pranhita-Godavari region. The Kamthi Formation overlies at a 10 to 15° angular unconformity the Motur and Barakar Formations.

Wardha Valley. The data (Fig. 5) are from Raja-Rao (1982, p. 66). "At places, the Kamthi Formation completely overlaps the Barakar Formation and rests over the Talchir sediments" (Raja-Rao, 1982, p. 65) at a disconformity or very low-angle unconformity. According to Dr. N. D. Mitra (personal communication, 1990), the Kamthi Formation, with *Daptocephalus* zone vertebrates, overlies at a low-angle unconformity the planed axis of a northwest-trending anticline in the Talchir and Barakar Formations. The reddish middle Kamthi Formation contains *Daptocephalus* zone vertebrates and *Glossopteris* below and, in the Mangli Bed above, vertebrates, including *Lystrosaurus*, as in the Panchet Formation (Satsangi, 1988, p. 248).

Northwest Pranhita-Godavari. The data (Fig. 5) are from Raja-Rao (1982, p. 11). The Kamthi Formation is divided into three parts: a lower part of coal measures and middle and upper parts of barren sandstone with redbeds. The reddish color starts in the latest Permian. Biostratigraphic ages, mainly by vertebrates, are from Kutty et al. (1988) and by palynology from Srivastava and Jha (1988). The Gangapur and Kota Formations overlap the Maleri Formation. Raja-Rao (1982, p. 14–15) reports that in a stream section west of Jaipuram the upper Kamthi overlaps all older formations, but this is not shown on his map.

Southeast Pranhita-Godavari. Lakshminarayana and Murty (1990) remapped the Chintalapudi Subbasin. They show that the Kamthi Formation, with *Glossopteris* in red claystone, which we identify as middle Kamthi (Fig. 5), disconformably overlies the Barakar Formation and in places is faulted against the Precambrian, contrary to Pande's (1988, p. 54) view that the Kamthi overlaps the Precambrian of the Mailaram High and of an outlier at 17°05′N, 80°36′E. The Barakar Formation is estimated to be 350 m thick (Vijayam and Deshpande, 1979, p. 646; Lakshminarayana and Murty, 1990).

Krishna-Godavari. The data (Fig. 5) are from Raja-Rao (1982, p. 11) and Prasad and Jain (1994), who describe a new Triassic section of 1,035 m of sandstone with minor shale unconformably between the Barakar Formation below and the Golapilli Sandstone above. The Cretaceous and Cenozoic section (Fig. 6, Table 3) is from S. P. Kumar (1983). Jaiprakash et al. (1993) date the the Deccan (Rajahmundry) Trap as Cretaceous/Tertiary (K/T) by its position between Maastrichtian foraminiferan-bearing shale below and Paleocene (P2) shale above, and Baksi et al. (1994) determined a $^{40}Ar/^{39}Ar$ age of 64.0 ± 0.4 Ma, marginally younger than the average age of 65.5 ± 0.5 Ma of the Deccan Trap but possibly coeval with lavas in the Kolhapur Formation of the Deccan Trap near the west coast.

Palar (located on Fig. 1; not plotted in Fig. 5). The data are from Raja-Rao (1982, p. 4) and Murthy and Ahmad (1979, p. 517, outcrop map on p. 521). The Talchir Formation, with leiospherids (Venkatachala and Tiwari, 1988, p. 26), rests on the Precambrian.

REFERENCES CITED

Acharyya, S. K., Shah, S. C., Ghosh, S. C., and Ghosh, R. N., 1979, Gondwana of the Himalaya and its biostratigraphy, in Laskar, B., and Raja-Rao, C. S., eds., Fourth International Gondwana Symposium: Delhi, India, Hindustan Publishing Corporation, p. 420–433.

Ahmad, F., 1966, Post-Gondwana faulting in Peninsular India and its bearing on the tectonics of the sub-continent: Aligarh, Aligarh Muslim University, Annals of the Geology Department, v. 2, 64 p.

Ahmad, N., 1975, Sone Valley Talchir glacial deposits, Madhya Pradesh, India: Journal of the Geological Society of India, v. 16, p. 475–484.

Ahmad, N., 1981, Late Palaeozoic Talchir tillites of Peninsular India, in Hambrey, M. J., and Harland, W. B., eds., Earth's pre-Pleistocene glacial record: Cambridge, Cambridge University Press, p. 326–330.

Ahmad, N., and Hashimi, N. H., 1974, New exposures of Talchir striated pavements in Madhya Pradesh: Current Science, v. 43, p. 614–615.

Ahmad, N., and Hashimi, N. H., 1976, Talchir glacial deposits of Hasdo and Gej Valleys in Madhya Pradesh: Quarterly Journal of the Geological, Mining and Metallurgical Society of India, v. 47, p. 143–149.

Ahmad, N., Ghauri, K. K., Abbas, S. M., and Moakhar, C. R., 1976, Basal Talchir striated pavements from lower Hasdo Valley, Madhya Pradesh, India: Indian Geological Association Bulletin, v. 9, p. 51–52.

Alam, M. K., Hasan, A.K.M.S., Khan, M. R., and Whitney, J. W., 1990, Geological map of Bangladesh: Dhaka, Geological Survey of Bangladesh, scale 1:1 000 000.

Anderson, J. M., 1981, World Permo-Triassic correlations: Their biostratigraphic basis, in Cresswell, M. M., and Vella, P., eds., Gondwana Five: Rotterdam, A. A. Balkema, p. 3–10.

Archbold, N. W., 1982, Correlation of the Early Permian faunas of Gondwana: Implications for the Gondwanan Carboniferous-Permian boundary: Journal of the Geological Society of Australia, v. 29, p. 267–276.

Aslam, M., Arora, M., and Tewari, R. C., 1991, Heavy mineral suite in the Barakar sandstone of Moher sub-basin, Singrauli coalfield, central India: Journal of the Geological Society of India, v. 38, p. 66–75.

Baksi, A. K., 1994, Geochronological studies on whole-rock basalts, Deccan Traps, India: Evaluation of the timing of volcanism relative to the K-T boundary: Earth and Planetary Science Letters, v. 121, p. 43–56.

Baksi, A. K., Barman, T. R., Paul, D. K., and Farrar, E., 1987, Widespread Early Cretaceous flood basalt volcanism in eastern India: Geochemical data from the Rajmahal-Bengal-Sylhet Traps: Chemical Geology, v. 63, p. 133–141.

Baksi, A. J., Byerly, G. R., Chan, L.-H., and Farrar, E., 1994, Intracanyon flows in the Deccan province, India? Case history of the Rajahmundry Traps: Geology, v. 22, p. 605–608.

Balasundarum, M. S., Ghosh, P. K., and Dutta, P. K., 1970, Panchet sedimentation and origin of red beds, in Haughton, S. H., ed., Second Gondwana Symposium: Pretoria, South Africa, Council for Scientific and Industrial Research, p. 293–301.

Balme, B. E., 1970, Palynology of Permian and Triassic strata in the Salt Range and Surghar Range, West Pakistan, in Kummel, B., and Teichert, C., eds., Stratigraphic boundary problems: Permian and Triassic of West Pakistan: University of Kansas Special Publication 4, p. 305–453.

Balme, B. E., 1980, Palynology and the Carboniferous-Permian boundary in Australia and other Gondwana continents: Palynology, v. 4, p. 43–55.

Banerjee, I., 1966, Turbidites in a glacial sequence—A study from the Talchir Formation, Raniganj Coalfield, India: Journal of Geology, v. 74, p. 593–606.

Banerjee, M., 1988, Karharbari: A formation or biozone: Lucknow, The Palaeobotanist, v. 36, p. 37–50.

Baud, A., Magaritz, M., and Holser, W. T., 1989, Permian-Triassic of the Tethys: Carbon isotope studies: Geologische Rundschau, v. 78, p. 649–677.

Bendapudi, V. R. M., 1994, On the evolution of the Pranhita-Godavari Gondwana Basin, Andhra Pradesh, India, in Mitra, N. D., Acharyya, S. K., Chandra, P. R., Ghosh, A., Ghosh, S., and Guha, P. K. S., eds., Ninth International Gondwana Symposium: Calcutta, India, Geological Survey of India, p. 100–101.

Bergmark, S. L., and Evans, P. R., 1987, Geological controls on reservoir quality of the northern Perth Basin: Australian Petroleum Exploration Association Journal, v. 27, p. 318–330.

Besems, R. E., and Schuurman, W. M. L., 1987, Palynostratigraphy of Late Paleozoic glacial deposits of the Arabian Peninsula with special reference to Oman: Palynology, v. 11, p. 37–53.

Bhalla, S. N., 1983, India, in Moullade, M., and Nairn, A. E. M., eds., The Phanerozoic geology of the World. 2: The Mesozoic: Amsterdam, Elsevier, p. 305–351.

Bhattacharya, N., 1981, Depositional patterns in limestones of the Kota Formation (Upper Gondwana), Andhra Pradesh, India, in Cresswell, M. M., and Vella, P., eds., Gondwana Five: Rotterdam, A. A. Balkema, p. 135–139.

Bint, A. N., and Helby, R., 1988, Upper Triassic palynofacies and environmental interpretations for the Rankin Trend, northern Carnarvon Basin, W.A., in Purcell, P. G., and Purcell, R. R., eds., The North West Shelf, Australia: Perth, Petroleum Exploration Society of Australia, p. 591–598.

Biswas, S. K., 1971, The Miliolite rocks of Kutch and Kathiawar (western India): Sedimentary Geology, v. 5, p. 147–164.

Biswas, S. K., 1981, Basin framework palaeoenvironment and depositional history of the Mesozoic sediments of the Kutch Basin, western India: Quarterly Journal of the Geological, Mining and Metallurgical Society of India, v. 53, p. 56–85.

Biswas, S. K., and Deshpande, S. V., 1983, Geology and hydrocarbon prospects of Kutch, Saurashtra and Narmada Basins: Petroleum Asia Journal, v. 6, p. 111–126.

BMR Palaeogeographic Group, 1990, Evolution of a continent: Canberra, Australian Bureau of Mineral Resources, 97 p.

Boast, J., and Nairn, A. E. M., 1982, An outline of the geology of Madagascar, in Nairn, A. E. M., and Stehli, F. G., eds., The ocean basins and margins. Volume 6: The Indian Ocean: New York, Plenum, p. 649–696.

Boote, D. R. B., and Kirk, R. B., 1989, Depositional wedge cycles on evolving plate margin, western and northwestern Australia: American Association of Petroleum Geologists Bulletin, v. 73, p. 216–243.

Bose, P. K., Mukhopadhyay, G., and Bhattacharyya, H. N., 1992, Glaciogenic coarse clastics in a Permo-Carboniferous bedrock trough in India: A sedimentary model: Sedimentary Geology, v. 76, p. 79–97.

Bose, U., and Ramanamurthy, B. V., 1979, Pattern of Talchir sedimentation in South India, in Laskar, B., and Raja-Rao, C. S., eds., Fourth International Gondwana Symposium: Delhi, India, Hindustan Publishing Corporation, p. 368–382.

Braakman, J. H., Levell, B. K., Martin, J. H., Potter, T. L., and van Vliet, A., 1982, Late Palaeozoic Gondwana glaciation in Oman: Nature, v. 299, p. 48–50.

Bradshaw, M. T., Yeates, A. N., Beynon, R. M., Brakel, A. T., Langford, R. P., Totterdell, J. M., and Yeung, M., 1988, Palaeogeographic evolution of the North West Shelf region, in Purcell, P. G., and Purcell, R. R., eds., The North West Shelf, Australia: Perth, Petroleum Exploration Society of Australia, p. 29–54.

Brookfield, M. E., 1993, The Himalayan passive margin from Precambrian to Cretaceous times: Sedimentary Geology, v. 84, p. 1–35.

Brookfield, M. E., and Sahni, A., 1987, Paleoenvironments of the Lameta Beds (late Cretaceous) of Jabalpur, Madhya Pradesh, India: Soils and biotas of a semi-arid alluvial plain: Cretaceous Research, v. 8, p. 1–14.

Campbell, I. H., Czamanske, G. K., Fedorenko, V. A., Hill, R. I., and Stepanov, V., 1992, Synchronism of the Siberian Traps and the Permian-Triassic boundary: Science, v. 258, p. 1760–1763.

Casshyap, S. M., 1973, Palaeocurrents and palaeogeographic reconstruction of the Barakar (Lower Gondwana) sandstones of Peninsular India: Sedimentary Geology, v. 9, p. 283–303.

Casshyap, S. M., 1979a, Patterns of sedimentation in Gondwana Basins, in Laskar, B., and Raja-Rao, C. S., eds., Fourth International Gondwana Symposium: Delhi, India, Hindustan Publishing Corporation, p. 525–551.

Casshyap, S. M., 1979b, Palaeocurrents and basin framework—An example from Jharia Coalfield, Bihar (India), in Laskar, B., and Raja-Rao, C. S., eds., Fourth International Gondwana Symposium: Delhi, India, Hindustan Publishing Corporation, p. 626–641.

Casshyap, S. M., 1981a, Lithofacies analysis of the Late Permian Raniganj coal measures (Mahuda Basin) and their palaeogeographic implications, in Cresswell, M. M., and Vella, P., eds., Gondwana Five: Rotterdam, A. A. Balkema, p. 125–133.

Casshyap, S. M., 1981b, Palaeodrainage and palaeogeography of the Son Valley Gondwana Basins, Madhya Pradesh, in Valdiya, K. S., ed., Geology of Vindhyanchal: Delhi, India, Hindustan Publishing Corporation, p. 132–142.

Casshyap, S. M., and Aslam, M., 1992, Deltaic and shoreline sedimentation in Saurashtra Basin, Western India: An example of infilling in an Early Cretaceous failed rift: Journal of Sedimentary Petrology, v. 62, p. 972–991.

Casshyap, S. M., and Kumar, A., 1987, Fluvial architecture of the Upper Permian Raniganj coal measure in the Damodar Basin, eastern India: Sedimentary Geology, v. 51, p. 181–213.

Casshyap, S. M., and Qidwai, H. A., 1971, Palaeocurrent analysis of Lower Gondwana sedimentary rocks, Pench Valley coalfield, Madhya Pradesh (India): Sedimentary Geology, v. 5, p. 135–146.

Casshyap, S. M., and Qidwai, H. A., 1974, Glacial sedimentation of Late Paleozoic Talchir Diamictite, Pench Valley Coalfield, Central India: Geological Society of America Bulletin, v. 85, p. 749–760.

Casshyap, S. M., and Srivastava, V. K., 1988, Glacial and proglacial sedimentation in Son-Mahanadi Gondwana basin: Paleogeographic reconstruction: American Geophysical Union, Geophysical Monograph 41, p. 167–182.

Casshyap, S. M., and Tewari, R. C., 1982, Facies analysis and paleogeographic implications of a late Paleozoic glacial outwash deposit, Bihar, India: Journal of Sedimentary Petrology, v. 52, p. 1243–1256.

Casshyap, S. M., and Tewari, R. C., 1984, Fluvial models of the Lower Permian coal measures of Son-Mahanadi and Koel-Damodar Valley basins, India: Special Publications of the International Association of Sedimentology, v. 7, p. 121–147.

Casshyap, S. M., and Tewari, R. C., 1988, Depositional model and tectonic evolution of Gondwana basins: Lucknow, The Palaeobotanist, v. 36, p. 59–66.

Casshyap, S. M., Tewari, R. C., and Khan, Z. A., 1988, Interrelationships of stratigraphic and lithologic variables in Permian fluviatile Gondwana coal measures of Eastern India, in Agterberg, F. P., and Rao, C. N., eds., Recent advances in quantitative stratigraphic correlation: Delhi, India, Hindustan Publishing Corporation, p. 111–125.

Casshyap, S. M., Tewari, R. C., and Khan, A., 1993a, Alluvial fan origin of the Bagra Formation (Mesozoic Gondwana) and tectono-stratigraphic implications: Journal of the Geological Society of India, v. 42, p. 269–279.

Casshyap, S. M., Tewari, R. C., and Srivastava, V. K., 1993b, Origin and evolution of intracratonic Gondwana basins and their depositional limits in relation to Son-Narmada Lineament, in Casshyap, S. M., Valdiya, K. S., Khain, V. E., Milanovsky, E. E., and Raza, M., eds., Rifted basins and aulacogens: Geological and geophysical approach: Nainital, India, Gyanodaya Prakashan, p. 200–215.

Chakrabarti, U., 1992, Intrabasinal differential and its influence on coal thickness configuration, Barakar Formation, Sohagpur coalfield, Madhya Pradesh: Journal of the Geological Society of India, v. 39, p. 212–222.

Chakrvarti, D. K., 1981, Gondwana stratigraphy: Review and revision: Quarterly Journal of the Geological, Mining and Metallurgical Society of India, v. 53, p. 125–153.

Chanda, S. K., 1968, Juras-Cretaceous stratigraphy and sedimentation around Jabalpur, Madhya Pradesh, and their paleogeographic implications: Journal of the Geological Society of India, v. 9, p. 21–31.

Chandra, S., 1991, Middle Gondwanas of Madhya Pradesh—A critical reappraisal: Journal of the Geological Society of India, v. 63, p. 229–245.

Chandra, S., and Betekhtina, O. A., 1990, Bivalves in the Indian Gondwana coal measures: Indian Journal of Geology, v. 62, p. 18–26.

Chandra, S. K., 1994, Marine signatures in the Gondwanas of peninsular India and Permian palaeogeography, in Mitra, N. D., Acharyya, S. K., Chandra, P. R., Ghosh, A., Ghosh, S., and Guha, P. K. S., eds., Ninth International Gondwana Symposium: Calcutta, India, Geological Survey of India, p. 61–62.

Chaudhuri, S., 1988, Marine influence in Hutar Coalfield, Bihar: Lucknow, The Palaeobotanist, v. 36, p. 30–36.

Chaudhuri, S., and Mondal, K., 1989, Eotomariidae gastropod in Talchir Formation of Ramgarh Coalfield and its paleoenvironmental significance: Journal of the Geological Society of India, v. 34, p. 665–668.

Chowdhury, S., 1990, Coccoliths in the Barakar Formation of Ramgarh coalfield, Bihar: Journal of the Geological Society of India, v. 35, p. 520–523.

Claoué-Long, J. C., Zichao, Z., Guogan, M., and Shaohua, D., 1991, The age of the Permian-Triassic boundary: Earth and Planetary Science Letters, v. 105, p. 182–190.

Conaghan, P. J., Shaw, S. E., and Veevers, J. J., 1994, Sedimentary evidence of the Permian/Triassic global crisis induced by the Siberian hotspot, in Beauchamp et al., eds., Pangea: Global environments and resources: Canadian Society of Petroleum Geologists Memoir 17, p. 785–795.

Das, D., Saha, T., and Bhattacharya, R., 1993, Sedimentological study of subsurface Gondwana sediments of Bengal Basin, in Dutta, K. K., and Sen, S., eds., Gondwana Geological Magazine Special Volume: Gondwana Geological Society, Department of Geology, Nagpur, India, p. 19–40.

Dasgupta, K., 1993, Some contributions to the stratigraphy of the Yerrapalli Formation, Pranhita-Godavari Valley, Deccan, India: Journal of the Geological Society of India, v. 42, p. 223–230.

Datta, A. K., 1983, Geological evolution and hydrocarbon prospects of Rajasthan Basin: Petroleum Asia Journal, v. 6, p. 93–100.

Datta, A. K., 1986a, Phosphate occurrences in Gondwana sediments: Geological Survey of India Coal Wing News, v. 6, p. 2.

Datta, A. K., 1986b, Map showing sedimentological thicknesses within Gondwana basin of peninsular India: Geological Survey of India Coal Wing News, v. 6, p. 24.

Datta, A. K., 1988, Talcher Coalfield: Geological Survey of India Coal Wing News, v. 8, p. 6.

Datta, N. R., and Mitra, N. D., 1984, Gondwana geology of Indian Plate—Its history of fragmentation and dispersion: Geological Survey of Japan Report 263, p. 1–25.

Datta, N. R., de-And, A. K., and Chakraborti, S. K., 1979, Environmental interpretation of Gondwana coal measures, in Laskar, B., and Raja-Rao, C. S., eds., Fourth International Gondwana Symposium: Delhi, India, Hindustan Publishing Corporation, p. 255–264.

De, A. K., 1979, Pebbly sandstone—A distinct horizon in some Gondwana coalfields in India, in Laskar, B., and Raja-Rao, C. S., eds., Fourth International Gondwana Symposium: Delhi, India, Hindustan Publishing Corporation, p. 697–708.

De, C., 1990, Upper Barakar Lebensspuren from Hazaribagh, India: Journal of the Geological Society of India, v. 36, p. 430–438.

De, C., 1993, Skolithos ichnofacies in the Barakar Formation from Hazaribagh, India and its fine resolution depositional and ecological significance: Gondwana Geological Magazine Special Volume, p. 512–521.

de Wit, M., Jeffery, M., Bergh, H., and Nicolaysen, L., 1988, Geological map of sectors of Gondwana reconstructed to their disposition ~150 Ma: Tulsa, Oklahoma, American Association of Petroleum Geologists, scale

1:10,000,000.

Dhar, P. C., and Singh, R. P., 1993, Evolution of Cambay Graben, in Casshyap, S. M., Valdiya, K. S., Khain, V. E., Milanovsky, M. M., and Raza, M., eds., Rifted basins and aulacogens: Geological and geophysical approach: Nainital, India, Gyanodaya Prakashan, p. 268–280.

Dickins, J. M., 1984, Late Palaeozoic glaciation: Australian Bureau of Mineral Resources Journal, v. 9, p. 163–169.

Dickins, J. M., and Shah, S. C., 1979, Correlation of the Permian marine sequence of India and Western Australia, in Laskar, B., and Raja-Rao, C. S., eds., Fourth International Gondwana Symposium: Delhi, India, Hindustan Publishing Corporation, p. 1–44.

Dickins, J. M., and Shah, S. C., 1981, Permian palaeogeography of Peninsular and Himalayan India and the relationship with the Tethyan region, in Cresswell, M. M., and Vella, P., eds., Gondwana Five: Rotterdam, A. A. Balkema, p. 79–83.

Dingle, R. V., Siesser, W. G., and Newton, A. R., 1983, Mesozoic and Tertiary geology of Southern Africa: Rotterdam, A. A. Balkema, 375 p.

Dolby, J., and Balme, B. E., 1976, Triassic palynology of the Carnarvon Basin, Western Australia: Reviews of Palaeobotany and Palynology, v. 22, p. 105–168.

Du Toit, A. L., 1937, Our wandering continents: Edinburgh, Scotland, Oliver and Boyd, 366 p.

Dutta, K. K., and Sen, S., eds., 1993, Scientific papers of Birbal Sahni Centenary National Symposium on Gondwana of India: Nagpur, Gondwana Geological Society, University Department of Geology, 527 p.

Dutta, P. K., and Ghosh, S. K., 1993, The century old problem of the Pali-Parsora-Tiki stratigraphy and its bearing on the Gondwana classification in peninsular India: Journal of the Geological Society of India, v. 42, p. 17–32.

Dutta, P. K., and Laha, C., 1979, Climatic and tectonic influence on Mesozoic sedimentation in Peninsular India, in Laskar, B., and Raja-Rao, C. S., eds., Fourth International Gondwana Symposium: Delhi, India, Hindustan Publishing Corporation, p. 563–580.

Dutta, P. K., and Mukherjee, K. N., 1979, Palaeo-slope control on differential deposition of coal in two adjacent sub-basins in the Singrauli Coalfield, India, in Laskar, B., and Raja-Rao, C. S., eds., Fourth International Gondwana Symposium: Delhi, India, Hindustan Publishing Corporation, p. 278–285.

Dutta, P. K., and Suttner, L. J., 1986, Alluvial sandstone composition and paleoclimates. II: Authigenic mineralogy: Journal of Sedimentary Petrology, v. 56, p. 346–358.

Dutta, P. M., and Yadagiri, P., 1994, An Early Jurassic mammalian fauna from India, in Mitra, N. D., Acharyya, S. K., Chandra, P. R., Ghosh, A., Ghosh, S., and Guha, P. K. S., eds., Ninth International Gondwana Symposium: Calcutta, India, Geological Survey of India, p. 25–26.

Elliot, D. H., 1975, Gondwana basins of Antarctica, in Campbell, K. S. W., ed., Gondwana geology: Canberra, Australian National University Press, p. 493–536.

Etheridge, M. A., and O'Brien, G. W., 1994, Structural and tectonic evolution of the Western Australian margin rift system: Australian Petroleum Exploration Association Journal, v. 34, p. 906–908.

Exon, N. F., Haq, B. U., and von Rad, U., 1992, Exmouth Plateau revisited: Scientific drilling and geological framework, in von Rad, U., and Haq, B. U., eds., Proceedings of the Ocean Drilling Program, Scientific Results, v. 122, p. 3–20.

Ferguson, K. U., 1981, Fission-track dating of shield areas, Australia: Relationships between tectonic and thermal histories and fission-track distributions [M.Sc. thesis]: Melbourne, University of Melbourne, 194 p.

Findlay, R. H., Unrug, R., Banks, M. R., and Veevers, J. J., eds., 1993, Gondwana Eight: Proceedings of the Eighth Gondwana Symposium: Rotterdam, A. A. Balkema, 623 p.

Foster, C. B., 1979, Permian plant microfossils of the Blair Athol Coal Measures, Baralaba Coal Measures, and the basal Rewan Formation of Queensland: Queensland Department of Mines, Geological Survey of Queensland Publication 372, 244 p.

Foster, C. B., 1982, Spore-pollen assemblages of the Bowen Basin, Queensland (Australia): Their relationship to the Permian/Triassic boundary: Review of Palaeobotany and Palynology, v. 36, p. 65–183.

Foster, C. B., and Waterhouse, J. B., 1988, The *Granulatosporites confluens* Oppel-zone and Early Permian marine faunas from the Grant Formation on the Barbwire Terrace, Canning Basin, Western Australia: Australian Journal of Earth Sciences, v. 35, p. 135–157.

Foster, C. B., Balme, B. E., and Helby, R., 1994, First record of Tethyan palynomorphs from the Late Triassic of East Antarctica: Australian Geological Survey Organisation, Journal of Australian Geology and Geophysics, v. 15, p. 239–246.

Fox, C. S., 1934, The Lower Gondwana coalfields of India: Geological Survey of India Memoir 59, xliv + 388 p. and 14 plates.

Frakes, L. A., and Crowell, J. C., 1970, Late Paleozoic glaciation. Part II: Africa exclusive of the Karroo Basin: Geological Society of America Bulletin, v. 81, p. 2261–2286.

Frakes, L. A., Kemp, E. M., and Crowell, J. C., 1975, Late Paleozoic Glaciation. Part VI: Asia: Geological Society of America Bulletin, v. 86, p. 454–464.

Gee, E. R., 1932, The geology and coal resources of the Raniganj Coalfield: Geological Survey of India Memoir 61, xiii + 343 p. and 20 plates.

Geological Survey of Western Australia, 1990, Geology and mineral resources of Western Australia: Western Australia Geological Survey Memoir 3, 827 p.

Ghosh, P. K., and Mitra, N. D., 1970a, A review of recent progress in the studies of the Gondwanas of India, in Haughton, S. H., ed., Second Gondwana Symposium: Pretoria, South Africa, Council for Scientific and Industrial Research, p. 29–47.

Ghosh, P. K., and Mitra, N. D., 1970b, Sedimentary framework of glacial and periglacial deposits of the Talchir of India, in Haughton, S. H., ed., Second Gondwana Symposium: Pretoria, South Africa, Council for Scientific and Industrial Research, p. 213–223.

Ghosh, P. K., and Mitra, N. D., 1975, History of Talchir sedimentation in Damodar Valley basins: Geological Survey of India Memoir 105, 117 p., 37 plates.

Ghosh, R., 1984, Dimensional fabric in Barakar Sandstones, Salanpur, West Bengal—An effective alternative for directional attributes: Quarterly Journal of the Geological, Mining and Metallurgical Society of India, v. 56, p. 95–100.

Ghosh, S. K., and Mukhopadhyay, A., 1985, Tectonic history of the Jharia Basin—An intracratonic Gondwana basin in Eastern India: Quarterly Journal of the Geological, Mining and Metallurgical Society of India, v. 57, p. 33–58.

Gleadow, A. J. W., and Duddy, I. R., 1984, Fission track dating of dolerite intrusions and widespread thermal effects on Permo-Carboniferous sediments in the northwestern Canning Basin, in Purcell, P. G., and Purcell, R. R., eds., The North West Shelf, Australia: Perth, Petroleum Exploration Society of Australia, p. 389–399.

Guha, P. K. S., Mukhopadhyay, S. K., and Das, R. N., 1994, Trace fossils as indicator of palaeoenvironment of Talchir Formation in Raniganj and Deogarh Group of coalfields, India, in Mitra, N. D., Acharyya, S. K., Chandra, P. R., Ghosh, A., Ghosh, S., and Guha, P. K. S., eds., Ninth International Gondwana Symposium: Calcutta, India, Geological Survey of India, p. 55–56.

Hälbich, I. W., Fitch, F. J., and Miller, J. A., 1983, Dating the Cape orogeny, in Söhnge, A. P. G., and Hälbich, I. W., eds., Geodynamics of the Cape Fold Belt: Geological Society of South Africa Special Publication 12, p. 149–164.

Haq, B. U., Hardenbol, J., and Vail, P. R., 1988, Mesozoic and Cenozoic chronostratigraphy and cycles of sea-level change, in Wilgus, C. K., Hastings, B. S., Kendall, G. C. St. C., Posamentier, H. W., Ross, C. A., and Van Wagoner, J. C., eds., Sea-level changes: An integrated approach: Society of Economic Paleontologists and Mineralogists Spe-

cial Publication 42, p. 71–108.
Helby, R., Morgan, R., and Partridge, A. D., 1987, A palynological zonation of the Australian Mesozoic: Association of Australasian Palaeontologists Memoir 4, p. 1–94.
Honegger, K., Dietrich, V., Frank, W., Gansser, A., Thöni, M., and Trommsdorff, V., 1982, Magmatism and metamorphism in the Ladakh Himalayas (the Indus-Tsangpo suture zone): Earth and Planetary Science Letters, v. 60, p. 253–292.
Jaeger, J. J., Courtillot, V., and Tapponnier, P., 1989, Paleontological view of the ages of the Deccan Traps, the Cretaceous/Tertiary boundary, and the India-Asia collision: Geology, v. 17, p. 316–319.
Jafar, S. A., 1982, Nannoplankton evidence of Turonian transgression along Narmada Valley, India, and Turonian-Coniacian boundary problem: Journal of the Palaeontological Society of India, v. 27, p. 17–30.
Jagannathan, C. R., Ratnam, C., Baishya, N. C., and Das-Gupta, U., 1983, Geology of the Offshore Mahanadi Basin: Petroleum Asia Journal, v. 6, p. 101–104.
Jain, S. L., and Roychowdhury, T., 1988, Fossil vertebrates from the Pranhita-Godavari Valley (India) and their stratigraphic correlation, in McKenzie, G. D., ed., Gondwana Six: Stratigraphy, Sedimentology, and Paleontology: Washington, D.C., American Geophysical Union, Geophysical Monograph 41, p. 219–228.
Jaiprakash, B. C., Singh, J., and Raju, D. S. N., 1993, Foraminiferal events across K/T boundary and age of Deccan Trap volcanism in Palakollu area, Krishna-Godavari basin, India: Journal of the Geological Society of India, v. 41, p. 105–118.
Johnson, B. D., Powell, C. McA., and Veevers, J. J., 1980, Early spreading history of the Indian Ocean between India and Australia: Earth and Planetary Science Letters, v. 47, p. 131–143.
Jowett, A., 1929, On the geological structure of the Karanpura Coalfields, Bihar and Orissa: Geological Survey of India Memoir 62, p. 1–144, plates 1–14.
Kapoor, H. M., 1992, Permo-Triassic boundary of the Indian subcontinent and its international correlation, in Sweet, W. C., Zunyi, Y., Dickins, J. M., and Hongfu, Y., eds., Permo-Triassic events in the Eastern Tethys: Stratigraphy, classification, and relations with the Western Tethys: Cambridge, Cambridge University Press, p. 21–35.
Kapoor, H. M., and Tokuoka, T., 1985, Sedimentary facies of the Permian and Triassic of the Himalayas, in Nakazawa, K., and Dickins, J. M., eds., The Tethys: Her paleogeography and paleobiogeography from Paleozoic to Mesozoic: Tokyo, Japan, Tokai University Press, p. 23–58.
Keating, B. H., and Sakai, H., 1991, Amery Group red beds in Prydz Bay, Antarctica: Proceedings of the Ocean Drilling Program, Scientific Results, v. 119, p. 795–809.
Kemp, E. M., 1975, The palynology of Late Palaeozoic glacial deposits of Gondwanaland, in Campbell, K. S. W., ed., Gondwana geology: Canberra, Australian National University Press, p. 397–413.
Kemp, E. M., Balme, B. E., Helby, R. J., Kyle, R. A., Playford, G., and Price, P. L., 1977, Carboniferous and Permian palynostratigraphy in Australia and Antarctica: A review: Australian Bureau of Mineral Resources Journal, v. 2, p. 177–208.
Keyser, N., and Zawada, P. K., 1988, Two occurrences of ash-flow tuff from the Lower Beaufort Group in the Heilbron-Frankfort area, northern Orange Free State: South African Journal of Geology, v. 91, p. 509–521.
Khan, Z. A., 1984, Significance of grain size frequency data in interpreting depositional environment of the Permian Barakar Sandstone in Rajmahal Basin, Bihar: Journal of the Geological Society of India, v. 25, p. 456–465.
Khan, Z. A., 1987, Paleodrainage and paleochannel morphology of a Barakar river (Early Permian) in the Rajmahal Gondwana Basin, Bihar, India: Palaeogeography, Palaeoclimatology, Palaeoecology, v. 58, p. 235–247.
Khan, Z. A., and Casshyap, S. M., 1979, Palaeocurrent pattern in Gondwana coal measures of Brahmini Coalfield (Santhal Pargana), Bihar, and possible regional implications: Journal of the Geological Society of India, v. 20, p. 197–204.
Khan, Z. A., and Casshyap, S. M., 1982, Sedimentological synthesis of Permian fluviatile sediments of East Bokaro Basin, Bihar, India: Sedimentary Geology, v. 33, p. 111–128.
Kreuser, T., Wopfner, H., Kaaya, C. Z., Markwort, S., Semkiwa, P. M., and Aslanidis, P., 1990, Depositional evolution of Permo-Triassic basins in Tanzania with reference to their economic potential: Journal of African Earth Sciences, v. 10, p. 151–167.
Krishna, J., Singh, I. B., Howard, J. D., and Jafar, S. A., 1983, Implications of new data on Mesozoic rocks of Kachchh, western India: Nature, v. 305, p. 390–392.
Kumar, A., 1983, Raniganj sedimentation in Damodar Valley Coalfields of eastern India [Ph.D. thesis]: Aligarh, India, Aligarh Muslim University, 177 p.
Kumar, S. P., 1983, Geology and hydrocarbon prospects of Krishna Godavari and Cauvery Basins: Petroleum Asia Journal, v. 6, p. 57–65.
Kumar, S. P., and Bhandari, L. L., 1973, Palaeocurrent analysis of the Athgarh Sandstone (Upper Gondwana), Cuttack District, Orissa (India): Sedimentary Geology, v. 10, p. 61–75.
Kutty, T. S., Jain, S. L., and Roychowdhury, T., 1988, Gondwana sequence of the northern Pranhita-Godavari Valley: Its stratigraphy and vertebrate faunas: Lucknow, The Palaeobotanist, v. 36, p. 214–229.
Lakshminarayana, G., and Murty, K. S., 1990, Stratigraphy of the Gondwana formations in the Chintalapudi sub-basin, Godavari valley, Andhra Pradesh: Journal of the Geological Society of India, v. 36, p. 13–25.
Lakshminarayana, G., Murty, K. S., and Rama-Rao, M., 1991, Stratigraphy of the upper Gondwana sediments in the Krishna-Godavari tract, Andhra Pradesh: Journal of the Geological Society of India, v. 39, p. 39–49.
Laskar, B., 1979: Evolution of Gondwana coal basin, in Laskar, B., and Raja-Rao, C. S., eds., Fourth International Gondwana Symposium: Delhi, India, Hindustan Publishing Corporation, p. 223–237.
Le Maitre, R. W., 1975, Volcanic rocks from Edel No. 1 petroleum exploration well, offshore Carnarvon Basin, Western Australia: Journal of the Geological Society of Australia, v. 22, p. 167–174.
Levell, B. K., Braakman, J. H., and Rutten, K. W., 1988, Oil-bearing sediments of Gondwana glaciation in Oman: American Association of Petroleum Geologists Bulletin, v. 72, p. 775–796.
Maheshwari, H. K., and Banerji, J., 1975, Lower Triassic palynomorphs from the Maitur Formation, West Bengal, India: Palaeontographica, v. 152-B, p. 149–190.
Mahoney, J. J., MacDougall, J. D., Lugmair, G. W., and Gopalan, K., 1983, Kerguelen hotspot source for Rajmahal Traps and Ninetyeast Ridge?: Nature, v. 303, p. 385–389.
McKelvey, B. C., and Stephenson, N. C. N., 1990, A geological reconnaissance of the Radok Lake area, Amery Oasis, Prince Charles Mountains: Antarctic Science, v. 2, p. 53–66.
Mehta, D. R. S., 1956, A revision of the geology and coal resources of the Raniganj Coalfield: Geological Survey of India Memoir 84, 113 p., 5 plates.
Mishra, D. C., Gupta, S. B., Rao, M.B.S.V., Venkatarayudu, M., and Laxman, G., 1987, Godavari Basin—A geophysical study: Journal of the Geological Society of India, v. 30, p. 469–476.
Mishra, H. K., 1986, A comparative study of the petrology of Permian coals from India and Western Australia [Ph.D. thesis]: Wollongong, Australia, University of Wollongong, 610 p.
Mishra, H. K., 1991, A comparison of the petrology of some Permian coals of India and Western Australia, in Ulbrich, H., and Rocha Campos, A. C., eds., Gondwana Seven: Proceedings of the Seventh Gondwana Symposium: São Paulo, Instituto de Geociências, Universidade de São Paulo, p. 603–613.
Mishra, H. K., Chandra, T. K., and Verma, R. P., 1990, Petrology of some Permian coals of India: International Journal of Coal Geology, v. 16, p. 47–71.
Mitra, N. D., 1987, Structures and tectonics of Gondwana basins of Peninsular India, in Singh, R. M., ed., Proceedings of the National Seminar on

Coal Resources of India: Banaras, India, Department of Geology, Banaras Hindu University, p. 30–41.

Mitra, N. D., 1991, The sedimentary history of Lower Gondwana coal basins of Peninsular India, in Ulbrich, H., and Rocha Campos, A. C., eds., Gondwana Seven: Proceedings of the Seventh Gondwana Symposium: São Paulo, Instituto de Geosciências, Universidade de São Paulo, p. 273–288.

Mitra, N. D., 1993a, Tectonic history of Lower Gondwana coal basins of Peninsular India: A reappraisal, in Casshyap, S. M., Valdiya, K. S., Khain, V. E., Milanovsky, E. E., and Raza, M., eds., Rifted basins and aulacogens: Geological and geophysical approach: Nainital, India, Gyanodaya Prakashan, p. 216–221.

Mitra, N. D., 1993b, Stratigraphy of Pali-Parsora-Tiki formations of South Rewa Gondwana Basin and Permo-Triassic boundary problem: Gondwana Geological Magazine Special Volume, p. 41–48.

Mitra, N. D., and Raja-Rao, C. S., 1981, Recent advances in the study of the Gondwana stratigraphy of India, in Cresswell, M. M., and Vella, P., eds., Gondwana Five: Rotterdam, A. A. Balkema, p. 86.

Mitra, N. D., Laskar, B., and Basu, U. K., 1975, Pattern of clastic dispersal in the Lower Gondwana coalfields of Peninsular India, in Campbell, K. S. W., ed., Gondwana geology: Canberra, Australian National University Press, p. 53–59.

Mitra, N. D., Bandyopadhyay, S. K., and Basu, U. K., 1979a, Sedimentary framework of the Gondwana sequence of Eastern India and its bearing on Indo-Antarctic fit, in Laskar, B., and Raja-Rao, C. S., eds., Fourth International Gondwana Symposium: Delhi, India, Hindustan Publishing Corporation, p. 37–41.

Mitra, N. D., Bose, U., and Dutta, P. K., 1979b, The problems of classification of the Gondwana Succession of the Peninsular India, in Laskar, B., and Raja-Rao, C. S., eds., Fourth International Gondwana Symposium: Delhi, India, Hindustan Publishing Corporation, p. 463–469.

Mitra, N. D., Acharyya, S. K., Chandra, P. R., Ghosh, A., Ghosh, S., and Guha, P. K. S., eds., 1994, Abstracts Ninth international Gondwana Symposium: Calcutta, India, Geological Survey of India, 188 p. + i–x.

Morante, R., Veevers, J. J., Andrew, A. S., and Hamilton, P. J., 1994, Determination of the Permian-Triassic boundary in Australia from carbon isotope stratigraphy: Australian Petroleum Exploration Association Journal, v. 34, p. 330–336.

Mukhopadhayay, A., 1984, Inter-relationship of tectonism and sedimentation in the Jharia Basin: Journal of the Geological Society of India, v. 25, p. 445–455.

Mukhopadhayay, A., Chaudhuri, P. N., and Banerji, A. L., 1984, Contemporaneous intrabasinal faulting in Gondwana Basin—The Jurabaga fault of Ib River Coalfield, a type example: Journal of the Geological Society of India, v. 25, p. 557–563.

Mukhopadhyay, D., Verma, R. P., and Jain, R. K., 1994, Fault pattern in the Jharia Coalfield, Bihar, India, in Mitra, N. D., Acharyya, S. K., Chandra, P. R., Ghosh, A., Ghosh, S., and Guha, P. K. S., eds., Ninth International Gondwana Symposium: Calcutta, Geological Survey of India, p. 96–97.

Murris, R. J., 1980, Middle East: Stratigraphic evolution and oil habitat: American Association of Petroleum Geologists Bulletin, v. 64, p. 597–618.

Murthy, N. G. K., and Ahmad, M., 1979, Paleogeographic significance of the Talchirs in the Palar Basin near Madras, India, in Laskar, B., and Raja-Rao, C. S., eds., Fourth International Gondwana Symposium: Delhi, India, Hindustan Publishing Corporation, p. 515–521.

Murti, K. S., and Lakshminarayana, G., 1994, Jurassic sedimentation in the Godavari Rift, India, in Mitra, N. D., Acharyya, S. K., Chandra, P. R., Ghosh, A., Ghosh, S., and Guha, P. K. S., eds., Ninth International Gondwana Symposium: Calcutta, Geological Survey of India, p. 107–108.

Naqvi, S. M., and Rogers, J. J. W., 1987, Precambrian geology of India: Oxford, Clarendon Press, 223 p.

Narayana-Murthy, B. R., and Hemmady, A. K. R., 1980, Barjora Coalfield: Geological Survey of India Memoir 88, p. 105–107.

Narayanan, V., 1977, Biozonation of the Uttatur Group, Trichinopoly, Cauvery Basin: Journal of the Geological Society of India, v. 18, p. 415–428.

Neogi, B. B., and Srivastava, U. C., 1987, A potential coal bearing belt under the Bengal alluvial plain, its structure and tectonics, West Bengal, India, in Singh, R. M., ed., Proceedings of the National Seminar on Coal Resources of India: Banaras, India, Department of Geology, Banaras Hindu University, p. 402–416.

Niyogi, B. N., 1987, Evolutionary scenario of Lower Gondwana coal basins of Peninsular India, in Singh, R. M., ed., Proceedings of the National Seminar on Coal Resources of India: Banaras, India, Department of Geology, Banaras Hindu University, p. 14–28.

Niyogi, D., 1961, Pattern of Talchir sedimentation in Burhai Gondwana Basin, Bihar, India: Journal of Sedimentary Petrology, v. 31, p. 63–71.

O'Brien, P. E., and Christie-Blick, N., 1992, Glacially grooved surfaces in the Grant Group, Grant Range, Canning Basin and the extent of Late Palaeozoic Pilbara ice sheets: Australian Bureau of Mineral Resources Journal, v. 13, p. 87–92.

Pakistani-Japanese Research Group, 1985, Permian and Triassic systems in the Salt Range and Surghar Range, Pakistan, in Nakazawa, K., and Dickins, J. M., eds., The Tethys: Her paleogeography and paleobiogeography from Paleozoic to Mesozoic: Tokyo, Japan, Tokai University Press, p. 221–311.

Pal, A. K., Sen, M. K., Ghosh, R. N., and Das, S. N., 1994, Marine incursions during Gondwana sedimentation in Damodar Valley basins, eastern India, in Mitra, N. D., Acharyya, S. K., Chandra, P. R., Ghosh, A., Ghosh, S., and Guha, P. K. S., eds., Ninth International Gondwana Symposium: Calcutta, India, Geological Survey of India, p. 63.

Palmer, A. R., 1983, The Decade of North American Geology 1983 geologic time scale: Geology, v. 11, p. 503–504.

Pande, B. C., 1988, Kamthi—A new concept: Lucknow, The Palaebotanist, v. 36, p. 51–57.

Pandya, K. L., 1990, Lithofacies and environment of deposition of a part of Talchir Group, Talchir Gondwana basin, Orissa: Journal of the Geological Society of India, v. 36, p. 175–186.

Pascoe, E. H., 1968, A manual of the geology of India and Burma, Volume 2: Calcutta, Government of India Press, p. 485–1338.

Peters, J., 1991, Structural framework and tectonic evolution of Satpura Basin: Madhya Pradesh: Oil and Natural Gas Commission Bulletin, v. 28, p. 889–908.

Powell, C. McA., and Li, Z. X., 1994, Reconstruction of the Panthalassan margin of Gondwanaland, in Veevers, J. J., and Powell, C. McA., eds., Permian-Triassic Pangean basins and foldbelts along the Panthalassan margin of Gondwanaland: Geological Society of America Memoir 184, p. 5–9.

Powell, C. McA., and Veevers, J. J., 1987, Namurian uplift in Australia and South America triggered the main Gondwanan glaciation: Nature, v. 326, p. 177–179.

Powell, C. McA., Roots, S. R., and Veevers, J. J., 1988, Pre-breakup continental extension in East Gondwanaland and the early opening of the eastern Indian Ocean: Tectonophysics, v. 155, p. 261–283.

Prasad, B., and Jain, A. K., 1994, Triassic palynoflora from the Krishna-Godavari Basin (India) and its stratigraphic significance: Journal of the Geological Society of India, v. 43, p. 239–254.

Prasada-Rao, R., 1976, A rational approach for exploraion of oil in the Cauvery and Godavari Basins, in Avasthi, D. N., and Varadarajan, K., eds., The workshop on coastal sedimentaries of India—South of 18°N latitude: General Proceedings: Madras, India, Oil and Natural Gas Commission, p. 31.

Price, P. L., Filatoff, J., Williams, A. J., Pickering, S. A., and Wood, G. R., 1985, Late Palaeozoic and Mesozoic palynostratigraphical units: CSR Oil and Gas Division, Palynology Facility, Report 274/25 (unpublished).

Rabu, D., Le Meteur, J., Bechennec, F., Beurrier, M., Villey, M., and Bourdillon-Jeudy de Grissac, C., 1990, Sedimentary aspects of the Eo-Alpine cycle on the northeast edge of the Arabian Platform (Oman Mountains), in Robertson, A. H. F., Searle, M. P., and Ries, A., eds., The geology and tectonics of the Oman region: Geological Society of London Special Publication 49, p. 49–68.

Radhakrishna, B. P., 1989, Welcome intensification of interest in Deccan Flood basalts: Journal of the Geological Society of India, v. 33, p. 285–290.

Raja-Rao, C. S., ed., 1981, Coalfields of India: Coalfields of North Eastern India: Geological Survey of India Bulletin 1, p. 1–26.

Raja-Rao, C. S., ed., 1982, Coalfields of India: Coal resources of Tamil Nadu, Andhra Pradesh, Orissa, and Maharashtra: Geological Survey of India Bulletin 2, 103 p.

Raja-Rao, C. S., ed., 1983, Coalfields of India: Coal resources of Madhya Pradesh, Jammu and Kashmir: Geological Survey of India Bulletin 3, 195 p.

Raja-Rao, C. S., ed., 1987, Coalfields of India: Coal resources of Bihar (excluding Dhanbad District): Geological Survey of India Bulletin 4, 336 p.

Raju, A. T. R., and Srinivasan, S., 1983, More hydrocarbons from well explored Cambay Basin: Petroleum Asia Journal, v. 6, p. 25–35.

Ramanamurthy, B. V., 1985, Gondwana sedimentation in Ramagundam—Mantheni area, Godavari Valley Basin: Journal of the Geological Society of India, v. 26, p. 43–55.

Ramanathan, S., 1981, Some aspects of Deccan volcanism of western Indian shelf and Cambay Basin: Geological Society of India Memoir 3, p. 198–217.

Ranga-Rao, A., Dhar, C. L., and Obergfell, F. A., 1979, Badhaura Formation of Rajasthan—Its stratigraphy and age, in Laskar, B., and Raja-Rao, C. S., eds., Fourth International Gondwana Symposium: Delhi, India, Hindustan Publishing Corporation, p. 481–490.

Rao, B. R. J., and Yadagiri, P., 1981, Cretaceous intertrappean beds from Andhra Pradesh and their stratigraphic significance: Geological Society of India Memoir 3, p. 287–291.

Rao, G. N., Satyanarayan, K., and Mohinuddin, S. K., 1993, Cretaceous sediments in the subsurface of Krishna-Godavari basin, India: Journal of the Geological Society of India, v. 41, p. 533–539.

Reddy, P. H., and Prasad, K. R., 1988, Paleocurrent and paleohydrologic analysis of Barakar and Kamthi Formations in the Manuguru Coalfield, Andhra Pradesh: Indian Journal of Earth Sciences, v. 15, p. 34–44.

Reeckmann, S. A., and Mebberson, A. J., 1984, Igneous intrusions in the northwest Canning Basin and their impact on oil generation, in Purcell, P. G., and Purcell, R. R., eds., The North West Shelf, Australia: Perth, Petroleum Exploration Society of Australia, p. 389–399.

Renne, P. R., and Basu, A. R., 1991, Rapid eruption of the Siberian Traps flood basalts at the Permo-Triassic boundary: Science, v. 253, p. 176–179.

Rishi, M. K., 1972, Gondwana palaeocurrents in Umaria, central India: Quarterly Journal of the Geological, Mining and Metallurgical Society of India, v. 47, p. 45–50.

Robertson, A. H. F., and Searle, M. P., 1990, The northern Oman Tethyan continental margin: Stratigraphy, structure, concepts and controversies, in Robertson, A. H. F., Searle, M. P., and Ries, A., eds., The geology and tectonics of the Oman region: Geological Society of London Special Publication 49, p. 3–25.

Robinson, P. L., 1967, The Indian Gondwana formations—A review, in Amos, A. J., ed., Gondwana stratigraphy, International Union of Geological Sciences Symposium, Buenos Aires, 1–15 October 1967: Paris, United Nations Educational, Scientific, and Cultural Organization, p. 201–268.

Roy, B. C., 1962, Geological map of India (sixth edition): Calcutta, Geological Survey of India, 1:2,000,000, secant conical projection, 11 sheets.

Roybarman, A., 1983, Geology and hydrocarbon prospects of west Bengal: Petroleum Asia Journal, v. 6, p. 51–56.

Rudra, D. K., 1982, Upper Gondwana stratigraphy and sedimentation in the Pranhita—Godavari Valley, India: Quarterly Journal of the Geological, Mining and Metallurgical Society of India, v. 54, p. 56–79.

Rust, I. C., 1975, Tectonic and sedimentary framework of Gondwana basins in Southern Africa, in Campbell, K. S. W., ed., Gondwana geology: Canberra, Australian National University Press, p. 537–564.

Saha, S. N., Roy, A. K., Brahman, C. V., Sastry, C. B. K., and De, M. K., 1992, Geophysical exploration for coal bearing Gondwana basins in the states of West Bengal and Bihar in north-east India: Tectonophysics, v. 212, p. 173–192.

Sarkar, A., Paul, D. K., Balasubrahmanyan, M. N., and Sengupta, N. R., 1980, Lamprophyres from Indian Gondwanas—K-Ar ages and chemistry: Journal of the Geological Society of India, v. 21, p. 188–193.

Sastri, V. V., Sinha, R. N., Singh, G., and Murti, K. V. S., 1973, Stratigraphy and tectonics of sedimentary basins on east coast of Peninsular India: American Association of Petroleum Geologists Bulletin, v. 57, p. 655–678.

Sastri, V. V., Raju, A. T. R., Sinha, R. N., and Venkatachala, B. S., 1974, Evolution of the Mesozoic sedimentary basins on the east coast of India: Australian Petroleum Exploration Association Journal, v. 14, p. 29–41.

Sastri, V. V., Raju, A. T. R., Sinha, R. N., Venkatachala, B. S., and Banerji, R. K., 1977, Biostratigraphy and evolution of the Cauvery Basin, India: Journal of the Geological Society of India, v. 18, p. 355–377.

Sastry, M. V. A., Acharyya, S. K., Shah, S. C., Satsangi, P. P., Ghosh, S. C., Raha, P. K., Singh, G., and Ghosh, R. N., 1977, Stratigraphic lexicon of Gondwana formations of India: Geological Survey of India Miscellaneous Publication 36, 170 p.

Satsangi, P. P., 1988, Vertebrate faunas from the Indian Gondwana Sequence: Lucknow, The Palaeobotanist, v. 36, p. 245–253.

Saxena, S. K., 1963, The Pachmarhi sandstones—A statistical study of size analysis: Journal of the Geological Society of India, v. 4, p. 116–129.

Sen, D. P., and Banerji, T., 1991, Permo-Carboniferous proglacial-lake sedimentation in the Sahajuri Gondwana basin, India: Sedimentary Geology, v. 71, p. 47–58.

Sen, D. P., and Pradhan, S. N., 1992, Step-like delta from Talchir rocks around village Simipal, Orissa: Journal of the Geological Society of India, v. 40, p. 105–114.

Sen, D. P., and Sinha, T. C., 1985, Triassic aeolian sedimentation in the Auranga Gondwana Basin, Bihar, India: Sedimentary Geology, v. 43, p. 277–300.

Sen, K. K., Datta, R. K., and Bandyopadhaya, S. K., 1987, Birbhum Coalfield: A major coalfield discovered, in Singh, R. M., ed., Proceedings of the National Seminar on Coal Resources of India: Banaras, India, Department of Geology, Banaras Hindu University, p. 417–427.

Sengor, A. M. C., 1987, Tectonics of the Tethysides: Orogenic collage development in a collisional setting: Annual Review of Earth and Planetary Science, v. 15, p. 213–244.

Sengupta, D. P., 1992, *Metoposaurus maleriensis* Roychowdhury from the Tiki Formation of the Son-Mahanadi valley of central India: Indian Journal of Geology, v. 64, p. 300–305.

Sengupta, S., 1966, Geological and geophysical studies in western part of Bengal basin, India: Bulletin of the American Association of Petroleum Geologists, v. 50, p. 1001–1017.

Sengupta, S., 1970, Gondwana sedimentation around Bheemaram (Bhimaram), Pranhita-Godavari Valley, India: Journal of Sedimentary Petrology, v. 40, p. 140–171.

Singh, H. P., and Venkatachala, B. S., 1988, Lower Cretaceous spore-pollen assemblages in peninsular India: Lucknow, The Palaeobotanist, v. 36, p. 168–176.

Singh, I. B., 1981, Paleoenvironments and paleogeography of Lameta Group sediments (Late Cretaceous) in Jabalpur area, India: Journal of the Paleontological Society of India, v. 26, p. 28–53.

Singh, V. K., 1988, Structural investigations in coal exploration with special reference to the Indian coalfields: Minetech India, v. 9, p. 57–81.

Smith, A. J., 1963a, Evidence for a Talchir (lower Gondwana) glaciation: Striated pavement and boulder bed at Irai, central India: Journal of Sedimentary Petrology, v. 33, p. 739–750.

Smith, A. J., 1963b, A striated pavement beneath the basal Gondwana sediments on the Ajay River, Bihar, India: Nature, v. 198, p. 880.

Smith, G. C., and Cowley, R. G., 1987, The tectono-stratigraphy and petroleum potential of the northern Abrolhos Sub-basin, Western Australia: Australian Petroleum Exploration Association Journal, v. 27, p. 112–136.

Spring, L., Bussy, F., Vannay, J.-C., Huon, S., and Cosca, M. A., 1993, Early Permian granitic dykes of alkaline affinity in the Indian High Himalaya of Upper Lahul and SE Zanskar: Geochemical characterization and geotectonic implications, *in* Treloar, P. J., and Searle, M. P., eds., Himalayan tectonics: Geological Society of London Special Publication 74, p. 251–264.

Srinivasa-Rao, K., Sreenivasa-Rao, T., Raju, M. S., Ali-Khan, M. I., and Silekar, V. S., 1979, Gondwana sedimentation in the south-central part of the Godavari Valley, *in* Laskar, B., and Raja-Rao, C. S., eds., Fourth International Gondwana Symposium: Delhi, India, Hindustan Publishing Corporation, v. 1, p. 588–609.

Srivastava, S. C., 1988, Stratigraphic position and age of plant-bearing Nidpur beds: Lucknow, The Palaeobotanist, v. 36, p. 154–160.

Srivastava, S. C., and Jha, N., 1988, Palynology of Kamthi Formation in Godavari Graben: Lucknow, The Palaeobotanist, v. 36, p. 123–132.

Srivastava, S. C., Prakash, A., and Singh, T., 1988, Permian palynofossils from the eastern Himalaya and their genetic relationship: Lucknow, The Palaeobotanist, v. 36, p. 326–338.

Srivastava, V. K., 1970, Petrographic and fabric analysis of the basal Talchir diamictite (Upper Carboniferous) in the Damodar Valley coalfields and their bearing on the direction of sediment transport: Aligarh, Aligarh Muslim University, Abstracts of the International Symposium on the Stratigraphy and Mineral Resources of the Gondwana System, p. 15.

Stampfli, G., Marcoux, J., and Baud, A., 1991, Tethyan margins in space and time: Palaeogeography, Palaeoclimatology, Palaeoecology, v. 87, p. 373–409.

Sundaram, D., Maiti, A., and Singh, G., 1979, Upper Triassic mioflora from Tiki Formation of South Rewa Gondwana Basin, Madhya Pradesh, India, *in* Laskar, B., and Raja-Rao, C. S., eds., Fourth International Gondwana Symposium: Delhi, India, Hindustan Publishing Corporation, v. 1, p. 511–514.

Suttner, L. J., and Dutta, P. K., 1986, Alluvial sandstone composition and palaeoclimate. I: Framework mineralogy: Journal of Sedimentary Petrology, v. 56, p. 329–345.

Tewari, R. C., 1980, Lithofacies, sedimentary petrology and paleogeography of Gondwana lithic-fill of Giridih and adjoining basins of Bihar [Ph.D. thesis]: Aligarh, Aligarh Muslim University, 226 p.

Tewari, R. C., 1992, Upper Gondwana sedimentation in Damodar and Son basins of east-central India: Tectonic and paleogeographic implications: IX Convention (Abstracts), Pune, Indian Association of Sedimentologists, University of Poona, p. 10–11.

Tewari, R. C., 1995, Braided-fluvial depositional model of Late Triassic Gondwana (Mahadeva) rocks of Son Valley, central India: Tectonic and paleogeographic implications: Journal of the Geological Society of India, v. 45, p. 65–73.

Tewari, R. C., and Casshyap, S. M., 1978, Sediment transport direction in fluviatile Karharbari sandstone, Giridih basin, Bihar: Indian Journal of Earth Sciences, v. 5, p. 95–102.

Tewari, R. C., and Casshyap, S. M., 1982, Paleoflow analysis of Late Paleozoic Gondwana deposits of Giridih and adjoining basins and paleogeographic implications: Journal of the Geological Society of India, v. 23, p. 67–79.

Tewari, R. C., and Casshyap, S. M., 1983, Cyclicity in early Permian fluviatile Gondwana coal measures: An example from Giridih and Saharijuri Basins, Bihar, India: Sedimentary Geology, v. 35, p. 297–314.

Tewari, R. C., and Casshyap, S. M., 1994, Mesozoic tectonic and rifting events in peninsular India and their bearing on Gondwana stratigraphy and sedimentation, *in* Mitra, N. D., Acharyya, S. K., Chandra, P. R., Ghosh, A., Ghosh, S., and Guha, P. K. S., eds., Ninth International Gondwana Symposium: Calcutta, India, Geological Survey of India, p. 87.

Tewari, R. C., and Veevers, J. J., 1993, Gondwana basins of India occupy the middle of a 7500 km sector of radial valleys and lobes in central-eastern Gondwanaland, *in* Findlay, R. H., Unrug, R., Banks, M. R., and Veevers, J. J., eds., Gondwana Eight: Proceedings of the Eighth Gondwana Symposium: Rotterdam, A. A. Balkema, p. 507–512.

Tiwari, R. S., and Ram, A., 1986, Late Permian palynofossils from the Pali Formation, South Rewa Basin, Madhya Pradesh: Geological, Mining and Metallurgical Society of India Bulletin, v. 54, p. 250–255.

Tiwari, R. S., and Tripathi, A., 1987, *Dubrajisporites*—A new trilete reticulate miospore genus from the Late Triassic of the Rajmahal Basin, India: Alcheringa, v. 11, p. 139–149.

Tiwari, R. S., and Tripathi, A., 1988, Palynological zones and their climatic inference in the coal-bearing Gondwana of peninsular India: Lucknow, The Palaeobotanist, v. 36, p. 87–101.

Tiwari, R. S., and Vijaya, A., 1992, Permo-Triassic boundary on the Indian peninsula, *in* Sweet, W. C., Zunyi, Y., Dickins, J. M., and Hongfu, Y., eds., Permo-Triassic events in the Eastern Tethys: Stratigraphy, classification, and relations with the Western Tethys: Cambridge, Cambridge University Press, p. 37–45.

Truswell, E. M., 1980, Permo-Carboniferous palynology of Gondwanaland: Progress and problems in the decade to 1980: Australian Bureau of Mineral Resources Bulletin 5, p. 95–111.

Truswell, E. M., 1991, Data report: Palynology of sediments from Leg 119 drill sites in Prydz Bay, East Antarctica: Proceedings of the Ocean Drilling Program, Scientific Results, v. 119, p. 941–945.

Tumuluri, S. G., and Roychauduri, K. K., 1979, Indian coals, quality evaluation data: Dhanbad, Central Fuel Research Institute, v. 6, p. 237–243.

Turner, B. R., 1991, Depositional environment and petrography of preglacial continental sediments from Hole 740A, Prydz Bay, East Antarctica: Proceedings of the Ocean Drilling Program, Scientific Results, v. 119, p. 45–56.

Turner, B. R., and Padley, D., 1991, Lower Cretaceous coal-bearing sediments from Prydz Bay, East Antarctica: Proceedings of the Ocean Drilling Program, Scientific Results, v. 119, p. 57–60.

Veevers, J. J., ed., 1984, Phanerozoic earth history of Australia: Oxford, Clarendon Press, 418 p.

Veevers, J. J., 1988a, Morphotectonics of Australia's northwestern margin—A review, *in* Purcell, P. G., and Purcell, R. R., eds., The North West Shelf, Australia: Perth, Petroleum Exploration Society of Australia, p. 19–27.

Veevers, J. J., 1988b, Gondwana facies started when Gondwanaland merged in Pangea: Geology, v. 16, p. 732–734.

Veevers, J. J., 1989, Middle/Late Triassic (230±5 Ma) singularity in the stratigraphic and magmatic history of the Pangean heat anomaly: Geology, v. 17, p. 784–787.

Veevers, J. J., 1990a, Tectonic-climatic supercycle in the billion-year plate-tectonic eon: Permian Pangean icehouse alternates with Cretaceous dispersed-continents greenhouse: Sedimentary Geology, v. 68, p. 1–16.

Veevers, J. J., 1990b, Development of Australia's post-Carboniferous sedimentary basins: Petroleum Exploration Society of Australia Journal, v. 16, p. 25–32.

Veevers, J. J., 1993, Gondwana facies of the Pangean supersequence: A review, *in* Findlay, R. H., Unrug, R., Banks, M. R., and Veevers, J. J., eds., Gondwana Eight: Proceedings of the Eighth Gondwana Symposium: Rotterdam, A. A. Balkema, p. 513–520.

Veevers, J. J., 1994a, Pangea: Evolution of a supercontinent and its consequences for Earth's paleoclimate and sedimentary environments, *in* Klein, G. D., ed., Pangea: Paleoclimate, tectonics, and sedimentation during accretion, zenith, and breakup of a supercontinent: Geological Society of America Special Paper 288, p. 13–23.

Veevers, J. J., 1994b, Case for the Gamburtsev Subglacial Mountains of East Antarctica originating by mid-Carboniferous shortening of an intracratonic basin: Geology, v. 22, p. 593–596.

Veevers, J. J., and Powell, C. McA., 1979, Sedimentary-wedge progradation from transform-faulted continental rim: Southern Exmouth Plateau, Western Australia: American Association of Petroleum Geologists Bulletin, v. 63, p. 2088–2096.

Veevers, J. J., and Powell, C. McA., 1987, Late Paleozoic glacial episodes in Gondwanaland reflected in transgressive-regressive depositional sequences in Euramerica: Geological Society of America Bulletin, v. 98, p. 475–487.

Veevers, J. J., and Powell, C. McA., eds., 1994, Permian-Triassic Pangean basins and foldbelts along the Panthalassan margin of Gondwanaland: Geological Society of America Memoir 184, 368 p.

Veevers, J. J., and Tewari, R. C., 1995, Permian-Carboniferous and Permian-Triassic magmatism in the rift zone bordering the Tethyan margin of southern Pangea: Geology, v. 23, p. 467–470.

Veevers, J. J., Powell, C. McA., and Johnson, B. D., 1975, Greater India's place in Gondwanaland and in Asia: Earth and Planetary Science Letters, v. 27, p. 383–387.

Veevers, J. J., Powell, C. McA., and Roots, S. R., 1991, Review of seafloor spreading around Australia. I: Synthesis of the patterns of spreading: Australian Journal of Earth Sciences, v. 38, p. 373–389.

Veevers, J. J., Tewari, R. C., and Mishra, H. K., 1994a, Gondwana coal-bearing fan of the east-central Gondwanaland platform disrupted by Late Triassic–Jurassic rifting, in Mitra, N. D., Acharyya, S. K., Chandra, P. R., Ghosh, A., Ghosh, S., and Guha, P. K. S., eds., Ninth International Gondwana Symposium: Calcutta, India, Geological Survey of India, p. 71–72.

Veevers, J. J., Conaghan, P. J., and Shaw, S. E., 1994b, Turning point in Pangean environmental history at the Permian/Triassic (P/Tr) boundary, in Klein, G. D., ed., Pangea: Paleoclimate, tectonics, and sedimentation during accretion, zenith, and breakup of a supercontinent: Geological Society of America Special Paper 288, p. 187–196.

Veevers, J. J., Powell, C. McA., Collinson, J. W., and Lopez-Gamundi, O. R., 1994c, Synthesis, in Veevers, J. J., and Powell, C. McA., eds., Permian-Triassic Pangean basins and foldbelts along the Panthalassan margin of Gondwanaland: Geological Society of America Memoir 184, p. 331–353.

Veevers, J. J., Cole, D. I., and Cowan, E. J., 1994d, Karoo Basin and Cape Fold Belt, in Veevers, J. J., and Powell, C. McA., eds., Permian-Triassic Pangean basins and foldbelts along the Panthalassan margin of Gondwanaland: Geological Society of America Memoir 184, p. 223–279.

Veevers, J. J., Conaghan, P. J., and Powell, C. McA., 1994e, Eastern Australia, in Veevers, J. J., and Powell, C. McA., eds., Permian-Triassic Pangean basins and foldbelts along the Panthalassan margin of Gondwanaland: Geological Society of America Memoir 184, p. 11–171.

Veevers, J. J., Clare, A., and Wopfner, H., 1994f, Neocratonic magmatic-sedimentary basins of post-Variscan Europe and post-Kanimblan Eastern Australia generated by right-lateral transtension of Permo-Carboniferous Pangaea: Basin Research, v. 6, p. 141–157.

Venkatachala, B. S., and Sinha, R. N., 1986, Stratigraphy, age and paleoecology of Upper Gondwana equivalent of the Krishna-Godavari Basin, India: Lucknow, The Palaeobotanist, v. 35, p. 22–31.

Venkatachala, B. S., and Tiwari, R. S., 1988, Lower Gondwana marine incursions: Periods and pathways: Lucknow, The Palaeobotanist, v. 36, p. 24–29.

Vijayam, B. E., and Deshpande, Y. R., 1979, Lithofacies analysis of the Barakar Formation of Kothagudem area, Andhra Pradesh, India, in Laskar, B., and Raja-Rao, C. S., eds., Fourth International Gondwana Symposium: Delhi, India, Hindustan Publishing Corporation, v. 2, p. 642–648.

Visser, J. N. J., 1984, A review of the Stormberg Group and Drakensberg Volcanics in southern Africa: Palaeontologia Africana, v. 25, p. 5–27.

von Rad, U., Haq, B. U., Gradstein, F., Ludden, J., and the ODP Leg 122/123 shipboard scientific parties, 1989, Triassic to Cenozoic evolution of the NW Australian continental margin and the birth of the Indian Ocean (preliminary results of ODP Legs 122 and 123): Geologische Rundschau, v. 78, p. 1189–1210.

von Rad, U., Exon, N. F., Boyd, R., and Haq, B. U., 1992a, Mesozoic paleoenvironment of the rifted margin off NW Australia (ODP Legs 122/123): American Geophysical Union, Geophysical Monograph 70, p. 157–184.

von Rad, U., Haq, B. U., and 15 others, 1992b, Leg 122: Proceedings of the Ocean Drilling Program, Scientific Results, v. 122, 934 p.

Webb, J. A., and Fielding, C. R., 1993, Permo-Triassic sedimentation within the Lambert Graben, northern Prince Charles Mountains, East Antarctica, in Findlay, R. H., Unrug, R., Banks, M. R., and Veevers, J. J., eds., Gondwana Eight: Proceedings of the Eighth Gondwana Symposium: Rotterdam, A. A. Balkema, p. 357–369.

Wopfner, H., 1993, Structural development of Tanzanian Karoo basins and the break-up of Gondwana, in Findlay, R. H., Unrug, R., Banks, M. R., and Veevers, J. J., eds., Gondwana Eight: Proceedings of the Eighth Gondwana Symposium: Rotterdam, A. A. Balkema, p. 531–539.

Wright, R. P., and Askin, R. A., 1987, The Permian-Triassic boundary in the southern Morondava Basin of Madagascar as defined by plant microfossils, in McKenzie, G. D., ed., Gondwana Six: American Geophysical Union, Geophysical Monograph 41, p. 157–166.

Xiaochi, Jin, 1994, Sedimentary and paleogeographic significance of Permo-Carboniferous sequences in western Yunnan, China: Geologisches Institut der Universität zu Köln, Germany, Sonderveröffentlichungen, v. 99, 136 p.

Yemane, K., and Kelts, K., 1990, A short review of palaeoenvironments for Lower Beaufort (Upper Permian) Karoo sequences from southern to central Africa: A major Gondwana lacustrine episode: Journal of African Earth Sciences, v. 10, p. 169–185.

Zunyi, Y., Yuqi, C., and Hongzhen, W., 1986, The geology of China: Oxford, Clarendon Press, 303 p.

Zutsi, P. L., and Prabhakar, K. N., 1993, Evolution of southern part of Indian East Coast basins: Journal of the Geological Society of India, v. 41, p. 215–230.

MANUSCRIPT ACCEPTED BY THE SOCIETY DECEMBER 12, 1994

Index

[Italic page numbers indicate major references]

A

acritarchs, 15
Afghan block, 33, 52
Agglomeratic Slate, 35
alga, freshwater, 43
Alice Springs Orogeny, 33
alkaline basaltic flow, Ladakh, 43
alluvial fan, 23, 48
alluvium, 49
Amadeus Transverse Zone, 33
Amb Formation, 36, 57
Ambela granite, 36
Amery area, 36, 37
apatite, 40
Argo Abyssal Plain, 48
ash-flow tuff, 41
Athgarh basin, 11
Athgarh Formation, 23
Athgarh Sandstone, 21, 48, 49, 59

B

Badhaura area, 15
Badhaura Formation, 36, 58
Bagra Formation, 19, 21, 23, 48, 59
Bainmedart Coal Measures, 37
Bap, 11, 36
Bap Formation, 57, 58
Bap-Jaisalmer area, 58
Barakar Formation, 1, 3, *6*, 7, 11, 15, 16, 17, 18, 20, 57, 58, 59
Bareilly, 11
Barren Measures, 1, 3, *7*, 15, 16, 18, 36, 52, 57, 58, 59
Barrow Delta, 48
Bartenia communis, 43
basalt
    Narsapur area, 23
    Rajmahal area, 59
basin, defined, 2
Beaufort Group, 41, 42, 52
Beaver Lake, 37, 41, 43, 48
Bengal basin, 7, 23
Bhimaram Formation, 1
Bhimaram Sandstone, 7, 19
Bijawar Formation, 23
Bijawar Supergroup, 59
Bijori Formation, 7, 59
bivalves, 58
Bokaro, 15
Bokaro area, 48, 58
Bokaro basin, 15
Bokaro coalfield, 11, 58
Bonaparte basin, 39
Bowen basin, 49, 57
brachiopods
    Bap Formation, 57
    Talchir Formation, 57
breccia-conglomerate, Chikiala Formation, 23
Broome area, 39
Bunbury Basalt, 48, 49

C

Cambay basin, 31
Canning basin, 35, 39, 43, 52, 57
Cape Fold Belt, 37, 41, 42, 43, 52, 54
Cape Range Fracture Zone, 44
carbonaceous sediment, 54
carbonaceous shale
    Barakar Formation, 6
    Dubrajpur Formation, 59
carbonate reefs, 43
Cargonian upland, 35
Carnarvon basin, 11, 43
Chhattisgarh area, 31
Chhattisgarh upland, 9, 11, 15, 16, 19, 23, 35, 42, 48
Chhidru Formation, 38, 57
Chhidru Member, 57
Chikiala Formation, 7, 23, 48
Chikiala-Gangapur Formations, 21
Chinnur, 7
Chinnur inlier, 20, 44
Chintalapudi subbasin, 2, 59
Chotanagpur fan, 16, 19
Chotanagpur upland, 9, 11, 15, 16, 19, 34, 42
Chuperbhita Coalfield, 59
Cistecephalus zone, 57
Claraia shales, 42
clay
    Bhimaram Sandstone, 19
    Denwa Formation, 7
    Dharmaram Formation, 20
    Kota Formation, 20
    Maleri Formation, 20
    Pali Formation, 58
    Yerrapalli Formation, 19
claystone
    Denwa Formation, 19
    Kamthi Formation, 59
coal
    Barakar Formation, 6, 11, 15, 18
    basin, 36
    Beaver Lake, 52
    deposition, 1, 44, *52*
    Dubrajpur Formation, 20
    gap, 19, 44, *52*
    Karharbari Formation, 6, 11, 18
    Kota Formation, 20
    Lesueur Sandstone, 49
    measures, 1, 11, 15, 16, 36, 44, 54
    Pali Formation, 58
    Raniganj Formation, 17, 59
    Singrauli, 18
    thickness, 6
coccoliths, Barakar Formation, 6
Cockleshell Gully Formation, 49
Collie basin, 48, *49*
conglomerate
    Denwa Formation, 19
    Dubrajpur Formation, 20
    Kamthi Formation, 59
    Kota Formation, 7

conglomerate (continued)
    North Karanpura Coalfield, 17
    Panchet Formation, 19
    Talchir Formation, 6
Congo basin, 37
Congo-Kaokofeld upland, 35
Cooper basin, 49
corals, 37
crustal shortening, 33
Cuvier area, 48
Cyclolobus, 37
Cynognathus zone, 57

D

Daltonganj, 11, 58
Daltonganj-Deogarh-Rajmahal Hills, 17
Damodar area, 7, 19, 21, 23, 36
Damodar basins, 8, 11, 21, 23, 44, *49*
Damodar River, 7, 48, 49
Damodar valley, 17
Damodar-Singrauli valley, 18
Dampier basin, 38
Dandot Formation, 57
Daptocephalus zone, 57, 59
Darling Fault, 49
Deccan plateau, 32
Deccan Trap, 11, 15, 23, 49, 58, 59. *See also* Rajahmundry Traps
deformation, 8, 19, 20, 43, *44*, 49, 52
    Chotanagpur upland, 43
Denwa Formation, 7, 19, 23, 59
Deogarh, 58
Deogarh basin, 15, 23, 58
Deogarh-Rajmahal area, 19
deposition, 19, 20, 37, 49, 52
Derby area, 39
Dewanganj area, 7
Dharmaram Formation, 7, 20, 57
Dicroidium flora, 43
dinocyst, 43
dinoflagellates, 43
dolerite, 38, 40
    Bengal basin, 23
    Canning basin, 39
    Damodar basin, 23
    Deogarh basin, 23
    Mahanadi basin, 23
Domra, 16
downthrowing, 20
dropstones, Talchir Formation, 6
Dubrajpur Formation, 7, 17, 19, 20, 21, 44, 54, 57, 59

E

Eastern Ghat-Sukinda-Singhbhum thrust, 2
Ecca coal measures, 36, 52
Ecca Group, 37, 42
Edel-1 well, 38, 39, 42
Elliot Formation, 54
Enderby-1 well, 38, 39, 43

erosion
  Barakar Formation, 17
  Talchir Formation, 17
Eurydesma zone, 35
Exmouth Plateau, 43, 44, 56

F

Faizabad ridge, 11
Faizabad upland, 11, 15, 35, 36, 42
faulted basins, 34
faulting, 8, 49, *54*
  Chotanagpur upland, 43
  Koel-Damodar area, 1
  Perth Basin, 40
faults, 7, 21
  Barakar Formation, 20
  Damodar basins, 21, 44
  Kingori Sandstone, 40
  Koel-Damodar area, 19
  Nidpur Beds, 58
  North Karanpura area, 19
  Pranhita-Godavari area, 21, 23
  Rajmahal area, 19
  Raniganj basin, 16
  Talchir Formation, 20
  Wagina Sandstone, 38
fauna
  fish, 20
  marine invertebrate, 11
  Talchir Formation, 11
  Yerrapalli Formation, 57
feldspar, 8, 9
  Pachmarhi Sandstone, 19
fish fauna, Kota Formation, 20
Flagstone Bench Formation, 37, 41, 42, 43, 54
flexed basins, 34
flora, Raniganj, 58
folding, 8, 49
foraminiferans, 57, 58
fossils, 58
  trace, 6, 58
  See also palynomorphs

G

Gaik granite, 43
Gamburtsev Mountains, 33, 34, 42, *48*
Gamburtsev upland, 36, 37
Gangapur Formation, 23, 59
Garu Formation, 36
Gascoyne area, 48
glaciogenic sediment, 1, 35
Glossopteris, 59
gneiss, Barakar Formation, 11
Godavari area, 1
Godavari basin, 44, 49
Godavari River, 48
Golapilli Formation, 23
Golapilli Sandstone, 21, 48, 49, 59
Gondwanides foldbelt, 48
Gondwanides, 43, 44, *52*, 54
granite, 8
granite gneiss, 8

granitoids, 8, 36
Granulatisporites confluens zone, 57, 58
Guizhou region, South China, 41

H

heat, release, 1, 35, 52
heavy minerals, Barakar Formation, 6
Heilbron-Frankfort area, Orange Free State, 41
Himalaya-Tibet upland, drainage system, 48
Hirapur Formation, 57
Hoshangapad, 23
Hoshangabad upland, 23
hot spot, 23
Hutar, 11, 58
Hutar basin, 49

I

Ib River area, 7, 15, 16, 18, 59
Indian-Madagascar upland, 35
Indo-Australian rift zone, 33, 34, 35, 36
Infra-Trappean sediments, 23
invertebrates, 57, 58
  marine, 36, 57, 58
Ipswich, 44
Ipswich Coal Measures, 52, 54
Ironstone Shale, Barren Measures, 7, 58
Isalo Group, 43

J

Jabalpur, 9
Jabalpur Formation, 21, 23, 48, 58, 59
Jainti, 59
Jainty coalfield, 57
Jaipuram, 59
jasper
  Bijawar Formation, 23
  Denwa Formation, 19
  Kaimur Formation, 23
Jetty Member, Flagstone Bench Formation, 41
Jharia, 58
Jharia basin, 7, 11, 15, 21, 49
Jharia coalfield, 11, 15, 58
Jhingurdah Coal, 17
Jhingurdah Seam, 7, 58
Jeypore-Bastar, 2

K

Kaimur Formation, 23
Kamptee area, 7, 18, 59
Kamthi Formation, 1, 7, 17, 18, 57, 59
Karanpura basin, 15
Karharbari Formation, 1, 3, *6*, 11, 15, 18, 57, 58, 59
Karharbari-Barakar coal measures, 15
Karnakata, 2
Karoo basin, 35, 37, 48, 49

Karoo terrain, 49
Karoo-type basins, 34, 49
Kathwai Member, 57
Kerguelen, 23, 48
Kidodia coxi, 41
Kingori Sandstone, 40
Kockatea Formation, 39
Kockatea Shale, 38
Koel-Damodar area, 1, 7, 15, 19, 20, 54, 59
Koel-Damodar basins, 2, 9
Koel-Damodar valley, 16
Koel River, 48
Kolhapur Formation, 23, 59
Kommugudem-A well, 7, 19, 57
Korba, 9, 15
Korba area, 16
Korba coalfield, 11
Korba/Mandi-Raigarh area, 59
Kota Formation, 7, 20, 21, 23, 44, 54, 59
Krishna basin, 49
Krishna-Godavari area, 7, 19, 21, 59
Krishna-Godavari basin, 43, 57
Krishna River, 48
Krishna subbasin, 2
Kulti Formation, 58. *See also* Barren Measures

L

Ladakh, 43
Lambert Graben, 48
Lameta Beds, 58
Lameta Formation, 23, 59
lamprophyre
  Bengal basin, 23
  Damodar area, 23
  Damodar basin, 23
  Deogarh basin, 23
  Kockatea Formation, 39
  Mahanadi basin, 23
  Perth basin, 38
  Rajmahal Trap, 44
  Wittecara-1 well, 42
Leigh Creek, 44
leiosphaerids, 15, 57, 58
  Bap Formation, 58
  Barakar Formation, 58
  Ib River area, 16, 59
  Jharia basin, 11
  Palar basin, 11
  Rajmahal basin, 11
  Raniganj Formation, 17, 58
  Satpura basin, 11
  Talchir Formation, 58, 59
Lesueur Sandstone, 43, 49
Lhasa block, 33, 38, 52
limestone, Kota Formation, 20
Locker Shale, 39
Lower Beaufort Group, 37
Lower Sakamena Group, 37
Lugu Hill, 58
Lunatisporites pellucidus zone, 57
Lystrosaurus zone, 57, 59

M

Madagascar rift zone, 34, 36, 41
magmatism, 40
Mahadeva Formation, 58
Mahadeva Group, 59
Mahanadi area, 59
Mahanadi basin, 2, 8, 20, 23, 31, 32
Mahanadi River, 48
Mahanadi-Son basin, 9
Mahanadi valley, 11
Maihur, 11
Mailaram Arch, 2
Mailaram High, 21, 23, 48, 49, 59
Maitur Formation, 57
Maji-ja-Chumvi Formation, Kenya, 41, 42
Malakand granite, 36
Maleri Formation, 7, 20, 23, 57, 58, 59
Maluncha Hill, 11
Mandur area, 7
Manendragarh, 6, 11, 57
Manendragarh-Sohagpur area, 59
Mangli Beds, 57, 59
Mehjia, 16
metamorphic rocks, 7
Mianwali Formation, 57
microflora, 58, 59
microplankton, marine, 39
Mikumi Basin, 38
Minutosaccus crenulatus, 57
Mirkheri, 11
Mittiwali Member, 57
Moher subbasin, 7
Molteno Formation, 43, 44, 52, 54
Monghyr-Saharsa ridge, 15
Monghyr-Saharsa upland, 16, 17, 19, 36
Mongolian Plateau, drainage system, 48
Moolayember Formation, 52
Motur Formation, 7, 59
Mount Victoria Land, 48
Mount Victoria Land block, 43
mudstone
    Dharmaram Formation, 7
    Parsora Formation, 7, 21
    Supra-Panchet Formation, 7
Mysore, 11

N

Nagaur Formation, 58
Namchi, 36
Narmada lineament, 48
Narmada-Son lineament, 2
Narmada valley, 23
Narmia Member, 57
Narsapur, 23, 49
Natal Basin, 48
Nidpur Beds, 19, 58
Ninetyeast Ridge, 23, 48
Noonkanbah Formation, 40
North Karanpura area, 19, 58
North Karanpura basin, 7, 57
North Karanpura Coalfield, 17
Nyakaitu Basin, Tanzania, 38

O

Olpad fanglomerate, 31
Oman, 38
Omeishan Basalt, 41
Onslow Palynoflora, 43, 57
Outeniqua folding event, 41

P

Pachmarhi Sandstone, 7, 19, 23, 59
Palar, 59
paleocurrents, 20, 21
    Barakar Formation, 15
    Bhimaram Sandstone, 19
    Jabalpur Formation, 23
    Karharbari Formation, 15
    Lameta Formation, 23
    Maleri Formation, 15
    Panchet Formation, 19
    Pranhita-Godavari basin, 11, 16
    Raniganj basin, 16, 17
    Satpura basin, 16
    Son-Mahandi basin, 11
    Umaria, 11
    Yerrapalli Formation, 19
Paleotethys, 35
Pali Formation, 7, 18, 58
palynoflora, 58
    Kamthi Formation, 57
    Raniganj Formation, 57, 58
palynomorphs, 37, 57, 58, 59
Panchet Formation, 1, 3, 7, 18, *19*, 20, 57, 58, 59
Panjal Trap, 38, 52
Paraná basin, 49
Parbatpur dome, 7
Parsora Formation, 7, 20, 21, 58
Pearl-1 well, 40
peat, Barakar Formation, 6
Perindi-1 well, 40
Permotethys, 34, 35
Perth basin, 11, 38, 39, 43, 49
phonolite, Perth basin, 38
phosphate, Ib River, 59
phosphorite, 15
    Barren Measures, 7, 15, 16
    Raniganj area, 17
    Raniganj Formation, 7, 59
Phulbani, 11
Pilbara Block, 43, 44
Pilbara-Yilgarn block, 34
Playfordiaspora crenulata zone, 57
Poole Sandstone, 57
Pranhita River, 48
Pranhita-Godavari area, 3, 7, 17, 18, 20, 49, 59
Pranhita-Godavari basin, 2, 7, 8, 9, 11, 15, 16, 19, 20, 21, 23, 48, 57, 59
Pranhita-Godavari lobe, 2
Pranhita-Godavari-Satpura basin, 11
Pranhita-Godavari valley, 19

Productus, 37
Protohapoloxypinus microcorpus zone, 57
Protohapoloxypinus samoilovichii zone, 57
Prydz Bay, 42, 48
pumice lapilli, 41
pyrite, 15
pyroclastics, Ladakh, 43

Q

quartz, Pachmarhi Sandstone, 19
quartzite
    Bagra Formation, 23
    Barakar Formation, 11

R

Radok Conglomerate, 37
Rahum area, 7
Rajahmundry Traps, 23, 31, 49, 59. *See also* Deccan Trap
Rajasthan, 15
Rajhara, 58
Rajmahal area, 11, 20, 59
Rajmahal basin, 2, 15, 54
Rajmahal Formation, 7, 21
Rajmahal Hills, 7, 16, 17
Rajmahal Trap, 17, 23, 44, 48, 59
Ramgarh, 11
Raniganj area, 19, 48, 59
Raniganj basin, 11, 15, 16, 17, 57, 58
Raniganj Beds, 58
Raniganj coal measures, 15
Raniganj coalfield, 58
Raniganj Formation, 1, 3, 7, 8, *16*, 57, 58
Rankin area, 43
Rankin delta, 43
redbeds
    Beaver Lake, 41, 52
    deposition, 1
    Elliot Formation, 54
    Panchet Formation, 7, 18
    Prydz Bay, 42
    Raniganj Formation, 7
relaxation, 8, *9*, 23
reptiles, 58
Rewan Group, 57
rhyolite, 38, 39
    Dampier basin, 38
    Enderby-1 well, 43
rhythmite, Talchir Formation, 6
rifting, 8, *49*, 56
Ruhende Beds, 37

S

Saharsa-Monghyr upland, 37
Sahul Shoals-1 well, 59
Salt Range, 11, 15, 36, 38, 52, 57, 58
Samfrau Geosyncline, 52
sandstone
    Barakar Formation, 6, 7, 11

sandstone (continued)
   Barren Measures, 7
   Bhimaram Sandstone, 19
   Denwa Formation, 7
   Dharmaram Formation, 7, 20
   Dubrajpur Formation, 7, 20
   Isalo Group, 43
   Kamthi Formation, 59
   Karharbari Formation, 6, 11
   Kota Formation, 20, 44, 54
   Krishna-Godavari area, 19, 59
   Maleri Formation, 20
   Pali Formation, 58
   Panchet Formation, 7
   Parsora Formation, 7, 21
   Son area, 19
   Supra-Panchet Formation, 7, 20
   Talchir Formation, 6
   Yerrapalli Formation, 19
Sardi-Warchha Formation, 57
Satpura area, 7, 18, 19, 21, 59
Satpura basin, 9, 15, 16, 23
Satpura-Jabalpur region, 21
Satpura Range, 31
seafloor spreading, 48, 49, 52, 56
sedimentation, 8
shale
   Barakar Formation, 6, 7
   Dubrajpur Formation, 20
   Kamthi Formation, 59
   Karharbari Formation, 11
   Krishna-Godavari area, 19, 59
   Pali Formation, 58
   Panchet Formation, 7
   Son area, 19
   Supra-Panchet Formation, 7, 20
   Talchir Formation, 6
Shark Bay, 44
Siang, 36
Siberian Traps, 40, 41
Sibumasu block, 33, 52
Sikkim, 36
siltstone
   Barakar Formation, 6
   Barren Measures, 7
   Talchir Formation, 6
Singhbhum, 2
Singrauli, 15, 18, 58
Singrauli coalfield, 7, 17

Singrimari, 15, 59
Sohagpur area, 59
Somali Basin, 48
Son area, 7, 11, 19, 20
Son basin, 2, 19, 31
Son-Mahanadi area, 6
Son-Mahanadi basin, 11
Son-Umaria area, 20, 21
Son Valley area, 21, 44
South Rewa coalfield, 57, 58
Southern Boundary Fault, 49
stratigraphy, Barakar Formation, 7
Subansiri, 15
subsidence, 8, 11
sulfur content, Barakar coal, 15
Sullavai Formation, 7
Supra-Panchet Formation, 1, 3, 7, 19,
   20, 21, 43, 44, 54, 58, 59

T

Talcher, 9
Talcher area, 16, 20, 59
Talcher coalfield, 11, 15, 59
Talchir Formation, 1, 3, 6, 7, 9, 11, 17,
   20, 57, 58, 59
Tapinocephalus zone, 57
Tethyan sea, 11, 17, 42
Tigrisporites playfordii zone, 57
Tiki Beds, 58
Tiki Formation, 7
tillite, Talchir Formation, 6
Tobra Formation, 57
trace fossils, Barakar Formation, 6
trachyte, Perth basin, 38
Transantarctic Mountains, 54
Tredian Formation, 57
Tuli basin, 49
turbidites, Talchir Formation, 6

U

Umaria area, 7, 11, 15, 18, 57, 58
Umaria coalfield, 15
upift, 8, 19, 21, 23, 43, 48
   Chotanagpur upland, 43
   Deccan plateau, 32
   Rajmahal Hills, 16
   Son Valley, 44

upift (continued)
   thermal, 49
uranium, 40

V

Variscan collision, 33, 48, 52
vegetation, 19
vertebrates, 19, 20, 23, 57, 59
Vindhyan basin, 2
Vindhyan Range, 11, 23
Vindhyan upland, 23
Vohipanana-Ambatokapika limestone, 37
volcanics, 35, 36, 38, 39
   alkaline, 38
   synrift, 56
volcanism, 38, 39, 42
   pyroclastic, 41

W

WA-U3 unconformity, 38, 43
Wagina Sandstone, 38, 39
Wallaby-Perth Fracture Zone, 44
Wardha area, 11, 19, 23, 59
Wardha coalfield, 17, 57
Wardha-Kamptee area, 18
Wardha Valley, 57, 59
Wargal Formation, 57
Warnbro Group, 49
Warora coalfield, 15
West Bokaro, 11
Wittecarra-1 well, 39, 42
Witwatersrand Arch, 36
Wombat Plateau, 44
Woniusi Formation, 35
wrenching, 44

Y

Yarragadee Formation, 49
Yerrapalli Formation, 7, 17, 19, 57, 59
Yilgarn Block, 49
Yuman granite, 36
Yunnan, 36

Z

Zanskar, 38

Typeset in U.S.A. by Johnson Printing, Boulder, Colorado
Printed in U.S.A. by Malloy Lithographing, Inc., Ann Arbor, Michigan